FLEXIBILITY

Flexible Companies for the Uncertain World

Gill Eapen

CRC Press is an imprint of the
Taylor & Francis Group, an **informa** business

CRC Press
Taylor & Francis Group
6000 Broken Sound Parkway NW, Suite 300
Boca Raton, FL 33487-2742

First issued in paperback 2017

ISBN 13: 978-1-138-11239-1 (pbk)
ISBN 13: 978-1-4398-1632-5 (hbk)

Library of Congress Cataloging-in-Publication Data

Eapen, Gill.
Flexibility : flexible companies for the uncertain world / Gill Eapen.
p. cm.
Includes bibliographical references and index.
ISBN 978-1-4398-1632-5 (hbk. : alk. paper)
1. Organizational change. 2. Organizational effectiveness. 3. Management. 4. Business cycles. I. Title.

HD58.8.E175 2010
658.4'01--dc22 2009042045

Visit the Taylor & Francis Web site at
http://www.taylorandfrancis.com

and the CRC Press Web site at
http://www.crcpress.com

Contents

Preface

In physics, the Heisenberg uncertainty principle states that certain characteristics of a particle such as its position and momentum cannot be determined with precision at the same time. Uncertainty is the inability to precisely know an outcome in the present or in the future. Uncertainty exists not only in physics but also in every aspect of one's life—family, health, career, and many others. Flexibility allows individuals and organizations to deal with uncertainty better by identifying new information and incorporating it into decisions as soon as it becomes available (in real time). For an individual, it entails a variety of skills that are transferable and an environment that allows application of those skills without constraints. For an organization, flexibility (or lack of it) can be seen in its structure, its systems, and the strategies it deploys. It is fundamentally important in the management of all uncertainty faced by individuals, organizations, countries, and the world.

In this book, I explore both the existence of uncertainty and the use of flexibility to counteract it through ideas, stories, anecdotal evidence, and some speculation. This is not a technical book. It is based on my experiences in over four dozen companies in multiple industries and countries in the last two decades. I have seen companies succeed and fail due to ideas, management, or luck. In studying these cases, I also found a number of common attributes that may point to future success or failure for organizations. I have detailed these symptoms using a construct that embodies all aspects of organizations. It is applicable for all types of companies—large and small, old and young, private and public. You may find some of these ideas debatable, and you may disagree with some others. At the heart of this framework is a singular characteristic—flexibility—that provides individuals and organization with the ability to succeed in a world driven by uncertainty in every dimension.

There is much information and advice available from management experts on how to strategize, invest and divest, create business processes, implement systems, manage human resources, structure finance, and conduct marketing. In the last two decades, consulting companies have proliferated to address every need of companies. Investment banking firms stand ready to raise capital, make deals, and help companies succeed. However, none of the help provided by the experts has resulted in these companies succeeding when uncertainty dominates. Most companies seem to do well when the future is reasonably predictable and uncertainty is low. However, many fail spectacularly in a discontinuity—a sudden change in the environment. Many individuals and companies seem to have difficulty adapting to a shock or a heightened level of uncertainty—something they are not used to. It is in this context that I introduce the idea of flexibility—the ability to adapt to new information quickly. Many attempt to tackle uncertainty by trying to eliminate it through cleverly crafted hedges—mechanisms that allow them to remove uncertainty from one part of the business or the other. However, as we will see throughout this book, attempting to eliminate uncertainty is costly and reducing uncertainty in one part of the business may increase the overall risk faced by the company and make it vulnerable to shocks. Increasing flexibility is a much better way to manage and to take advantage of uncertainty, reducing risk and increasing economic value.

There is a body of work in the area of "decision making under uncertainty," including my previous book: *Decision Options: The Art and Science of Making Decisions*. There has been important academic research in this area, sometimes referred to as "real options" or "contingent claims analysis." Most of this work deals with the economic valuation of private assets such as private companies, patents, technologies, real estate, and anything that does not trade in the marketplace directly. The mathematics underlying the theory gets intimidating, but increasingly there are practical applications that are emerging to help companies understand economic value better and use that information in investment decision making. Academic literature in this area has demonstrated how a formal consideration of uncertainty and flexibility influences economic value and decisions, in stylized cases. However, making investment decisions, albeit important, is only one part of managing a company. Success depends not only on better decision making but also on creating and managing a flexible environment that allows the implementation of the decisions taken. Unless companies can implement the insights in a systematic way, they will not ultimately add much value to the company.

In this book, I investigate how companies can create an environment where implementation of options-based decisions is possible. Options-based decisions become important when uncertainty is present. It is, however, relevant only if there is sufficient flexibility that allows the company to take advantage of such uncertainty. I treat this in an informal fashion, and it does not require any technical background for the reader to appreciate it. The organization is analyzed holistically and its primary components—structure, system, and strategy—are treated independently and in detail. The book provides a framework to diagnose existing problems in a company and understand how to improve flexibility and reduce the risk of catastrophic failure. It is useful to analysts, managers, and decision makers in virtually any industry where uncertainty is present. It is also an introduction to students of business and strategy in conjunction with more formal work in economic valuation and portfolio management.

The book is organized as follows. The first chapter introduces flexibility at an individual level and then in companies. It discusses why flexibility is an effective tool in managing and taking advantage of uncertainty. In Chapter 2, I take a historical peek at the evolution of organizations and the environment we are currently in. Chapter 3 introduces the reader to the three primary components of an organization—structure, system, and strategy. In the next three chapters, Chapters 4–6, I delve into these three components in detail. Chapter 4 dissects structure into human, infrastructure, and information subcomponents. Chapter 5 explains the three major types of systems—technology, process, and content. In Chapter 6, we study strategy—internal strategies that focus on operations and external strategies that consider competition and the macro-economy. I also discuss boundary strategies that deal with partners and collaborators.

Chapter 7 details the metrics that can be used to measure flexibility in an organization. In Chapter 8, I conduct a flexibility audit for a hypothetical technology company and identify existing problems. Solutions are then suggested that will allow the company to increase flexibility and slowly transition away from its current rigid state. A project plan and specific actions are also provided. Chapter 9 shows how to design a flexible organization from scratch and the common pitfalls faced by emerging companies. Finally, in Chapter 10, I elevate the flexibility discussion to the level of countries and suggest better policies.

Flexibility is a holistic notion, and it touches many dimensions—individuals, companies, countries, and the world. I have attempted to touch on all dimensions, but the focus is primarily on organizations and how decision making and performance can be improved through flexibility. Even though the stories and case studies I have provided are from my experiences, all names of companies are fictional and the cases provided are stylized. I challenge the status quo ideas that were created during the Industrial Revolution and demonstrate why they have become irrelevant and even dangerous for contemporary organizations, mostly driven by information. If you are interested in better ways of structuring companies, creating systems, and managing strategy, you will find this book useful.

Author

Gill Eapen is the founder and managing principal of Decision Options, LLC—a boutique advisory services company. Decision Options® pioneered the practical application of options-based valuation of private assets to aid decision making in industries that show high levels of uncertainty and flexibility, such as life sciences, energy, technology, financial, and legal services. He conceived and led the development of a technology platform—Decision Options Technology—that allows modeling of complex decision problems and assets to improve decision making, risk management, and portfolio maximization.

Prior to establishing Decision Options, he was group director at Pfizer, responsible for the financial analysis and planning of the research and development (R&D) portfolio. During his tenure at Pfizer, Mr. Eapen conceived and led the development of a forecasting and capital allocation methodology and system that incorporate uncertainty emanating from all aspects of pharmaceutical R&D. He also conceived methodologies to incorporate flexibility in the management of uncertainty and helped senior decision makers in R&D to allocate resources more optimally across the portfolio. Before that, he was manager at Deloitte Consulting Group, providing advisory services to a variety of clients in software, technology, commodities, consumer goods, logistics, and manufacturing. Previous employers also include Hewlett Packard Company and Asea Brown Boveri.

Mr. Eapen holds graduate degrees from the University of Chicago and Northwestern University as well as an undergraduate degree from the Indian Institute of Technology. He is a CFA charter holder and a member of the Boston Security Analysts Society. His first book, *Decision Options: The Art and Science of Making Decisions*, details a formal treatment of economic valuation and decision making in companies. His blog, which focuses on market-based policy options, is titled "Ideas, Options, and Speculation." The details of Decision Options, LLC are available at http://decisionoptions.com.

1

Uncertainty and Flexibility

Today's world is embroiled in uncertainty at all levels—political, economic, social, and commercial. This is nothing new, although every generation seems to believe that they faced the worst. In the last hundred years, the world has seen much—world wars, nuclear bombs, airplanes, computers, and the Internet. Quantum theory was revealed, options theory was forged, planets were discovered outside the solar system, and the human genome was mapped. Bigger wars were followed by smaller but more potent ones, as some tried to spread philosophy and others, religion. Diseases spread across the globe by air, water, airplanes, and human contact but were conquered by chemicals and magic. Capital and labor clashed, some driven by profits and others by power, aided and abetted by philosophers of various persuasions. Countries were freed by war and peace, divided into many by walls and mountains, and then combined into one again. Democracy and religious extremism has spread hand in hand. Planned economies failed, while the foundations of the free market philosophy were also shaken. Companies grew to gigantic proportions, with revenues dwarfing the gross domestic product (GDP) of countries, and they spread across the globe, creating sprawling multinationals. They merged with others or acquired smaller rivals, creating even larger enterprises using size and revenue as the fundamental measure of success. Trade blocks and alliances formed, aided by geographical proximity of countries with different specializations, and then grew into unions of currencies and cultures. Bitter enemies became best friends and vice versa for oil and gold. Frozen alliances in a cold war broke into pieces while new ones were put together with no real objectives. Nobel laureates lamented about a warming globe fueled by dirty coal while dirtier energy policies went unchallenged.

The real economy, driven by companies that make tangible products and value-added services, was overtaken by the financial world where transactions were conceived in every form, aggregating and disaggregating pieces of ever diminishing productive assets. Companies formed on a song and a prayer raised many millions from capital providers, large and small, on the promise of yet-to-be-invented technologies. Financiers peddled securities of all kinds on the buy and the sell side and grew their operations so fast that intermediation became bigger than what was intermediated. Stock markets rose fast and then fell in days, only to rise again. *Leverage* became synonymous with *talent*, and mediocre ideas were levered up so as to manufacture hitherto unknown levels of returns. Then, in a momentary lapse of reason, some 100-year-old companies went bankrupt in less than a week in a spectacular show of incompetence, greed, and fraud. But some were saved by the intervention of policy makers, government bailouts have become as common as foreclosures, and we have reached the doors of a regime change with no precedence.

It is unclear whether we have just entered a period of upheaval and how long the current environment will continue. Financial markets continue melting down as investors digest the meaning of recent events and how they are going to affect asset prices in the short and in the long run. Anybody with a 401(k) knows well that the values of the financial assets, such as the stock of companies, have taken a nosedive. This was accompanied by a plunge in the value of real assets such as real estate. The spot and futures prices of commodities have also followed suit. The recent asset value deflation is unprecedented at many levels as it has affected every class of investment, financial and real, and has spread around the globe nearly instantly. Anybody watching the financial markets and the worldwide recession is keenly aware of the great difficulties individuals and companies are facing. What may not be obvious is that the volatility (the tendency for prices to move around) has moved up to unprecedented levels as well. This means that the asset prices did not move down smoothly but rather exhibited a high level of uncertainty day to day. Readers may be aware of the rapid rise and fall seen in the Dow Jones Industrial Average and the S&P 500 index—both representing an overall metric for the market. As fear and confusion gripped investors, their actions may have become irrational. Wild and inexplicable swings are happening in asset prices on a daily basis. Since the world economies are well integrated, the shocks in one part of the system spread quickly to other areas and often are amplified as investors panic due to lack of sufficient information. As an illustration, Figure 1.1 shows the chart of the volatility of the S&P 500 index (a combination of the largest 500 companies in the United States).

The y axis shows volatility, a measure of how much the prices move around from period to period. The more the prices move (up or down), the higher the volatility is. The x axis shows the month and year in which the volatility is calculated. For example, 12.5 means December 2005. As is evident from the chart, volatility was very low till about June 2007. It began to move higher slowly and this trend continued till September 2008. Then suddenly, we witnessed a dramatic and unprecedented rise in volatility—600% more than normal. Although it has shown some abatement since the beginning of 2009, it still remains two to three times more than what is typically seen in a broad index such as the S&P 500. Specific stocks or sector indexes may show volatilities many times more than what is represented here since the S&P 500 is a diversified and broad index. Any idea, business model, investment philosophy, and retirement plan based on status quo assumptions may have failed in this regime of catastrophic value loss combined with hypervolatility. This is a constant reminder to us that one cannot forecast too far into the future or

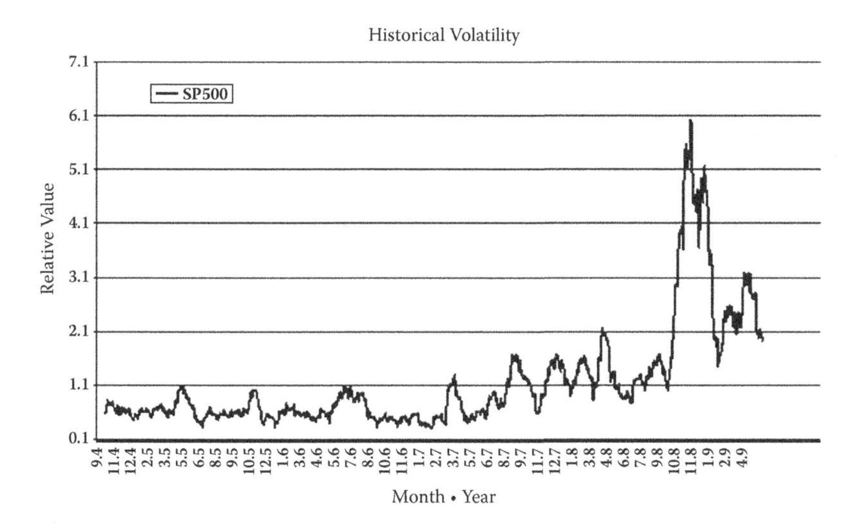

FIGURE 1.1
Volatility of the S&P 500 for the past several years.

plan with assumptions that remain constant over time. It is not only that the future is uncertain but also that the uncertainty may be manyfolds higher than what we are used to for the foreseeable future.

It is, indeed, a world of great uncertainty, opportunity, and peril. Uncertainty is the inability to precisely know an outcome in the future. Flexibility allows individuals and organizations to deal with uncertainty better. It allows individuals and organizations to identify new information and incorporate it into decisions as soon as it becomes available. For an individual, it entails a variety of skills that are transferable and an environment that allows application of those skills without constraints. For an organization, it may be seen in its structure, systems, and strategies. Flexibility is fundamentally important in the management of all uncertainty faced by individuals, organizations, countries, and the world.

Ever since the first bacterium formed in the primordial soup, flexibility has been the key to survival for organisms. They incorporated new information from the environment, survived through uncertain geological times, and became the most successful biological entity then and remain so now. Mutation allowed earlier organisms to randomly introduce flexibility into their structures and evolve into more complex systems. Humans, with an open mind and curiosity, rose out of the African savannahs, explored far and wide, accumulated knowledge, adapted, and became successful. In the last few thousand years, large and complex civilizations formed and failed, demonstrating how complex organizations can lose flexibility and die. Industrial enterprises of the past and the service behemoths of today demonstrate flexibility in varying levels. It is my belief that the primary differentiating factor between successful and failed companies is the flexibility existing in their structure, systems, and strategies. In the current financial meltdown and associated market chaos, many companies have failed and are failing because they subscribe to conventional designs and ideas. Many of these ideas have been with us since the Industrial Revolution. The world has changed much since then but not the organizations that inhabit it. In an information and knowledge driven environment that shows high levels of uncertainty, most common ideas of structure, system, and strategy are ineffective.

Uncertainty is on the increase in our personal lives as well. Most believe uncertainty is never good and consider uncertainty to be a synonym for risk that should always be avoided. Gone are the days where each of us could bank on a lifetime job with a large company, retirement, and a pleasant life with a provided pension. Gone are the days where we could settle down in a community and hope to be there for generations. As companies and businesses change and in many cases disintegrate into smaller units, driven by changes in technology and a closer integration of the world economy, individuals have to manage their lives and careers very differently from what we are used to. The primary reason for companies to exist, as Nobel Laureate Ronald Coase from the University of Chicago hypothesized in 1937, is the ability to decrease transaction costs. Transaction costs are the time and effort needed to transact with an external party. For example, if an automobile manufacturer contracts with suppliers for the components of the car it is making, it will incur significant time and effort to create and manage all those contracts. By vertically integrating—that is, making everything internally—the company can avoid such transactions and associated time, effort, and costs. Thus, we should see bigger and bigger companies as they backward integrate into raw materials and components and forward integrate into finished goods, so as to avoid all transaction costs. This, indeed, happened in the Industrial Revolution and much of the era following it.

The need for lowering transaction costs, however, has become moot in the modern world of low or no transaction costs. In today's world, Internet and other technologies make transactions virtually free. In fact, companies find themselves much better off focusing on exactly what they are good at and contract with other companies for everything else they need. As there can be many suppliers for the specific products and services that a company may be seeking, the cost of acquiring it from outside (especially if the service or product is not their core business) may turn out be much cheaper than making it internally. Since the suppliers can come from anywhere in the integrated world economy, going outside also helps the company take advantage of currency and labor cost differentials. As readers may be aware, many low cost countries (LCCs) have emerged, driven by large pools of talent (and associated lower labor costs) and artificially low currencies (due to the policies pursued by their governments). For example, China provides low-cost manufacturing and India specializes in low-cost services. For companies in the United States and the European Union (EU), making certain products and providing certain services internally will be prohibitively expensive compared to suppliers in these low cost countries. This dual advantage—lower search and transaction costs due to the Internet and the possibility of lower costs outside—actually forces companies in the modern era to be smaller and smaller. At the extreme, the most efficient company will be an individual. This means that each individual is now exposed to all the uncertainties companies are exposed to but also gains the opportunity to take advantage of them.

To succeed in the presence of high uncertainty, individuals have to increase flexibility through education, skills, language, and other attributes that span a variety of industries and outcomes. This means that most notions of stability—through lifetime employment and permanent home location—are not useful for the individual. If these still exist and the individual prefers them, the outcome is likely suboptimal. For example, the perceived safety of employment with a large company is a myth now as many companies are forced to change employment levels quickly—either by moving to different skill sets or to different locations. If companies are getting smaller due to the lower transaction and production costs for noncore items, employment in certain type of companies may cease to exist in the long run. For an individual, investing in oneself—through education and skills building—is the single most important activity that enhances flexibility and improves the individual's ability to manage uncertainty. Financial flexibility—the presence of low debt and access to cash—are equally important as this allows the individual to pursue the right opportunities as they become evident. Keeping fit and free of disease are also important as this removes constraints on a person's ability to take advantage of opportunities. Ability to converse in a variety of languages and to use many communication modalities also helps one to keep all options open. Communication modalities include the computer and various protocols provided in it. "Communication" thus has to be considered more generally and languages need to include both computer languages as well as specialized jargon created by widely used computer protocols such as "texting" and "twittering." Having an open mind and not having set biases against others also enhances the flexibility of the individual. Biases held by individuals against countries, languages, and personal attributes may reduce the choices available to them. This becomes increasingly critical as the scope of the transactions the individual will enter, encompass the world. As the individual is forced to transact with buyers and suppliers from all around the world, the knowledge of varied cultures and customs as well as the ability to communicate with a disparate group of people will become important advantages.

The need for flexibility has implications for those who are managing their careers through employment as well. The conventional wisdom of specializing in a single job and working one's way up through titles may not provide any flexibility to the individual if such a career track does not add skills to the portfolio. A better strategy would be rotating through various functions in a company to increase one's knowledge of various aspects. Skills that are transferable to other jobs add more flexibility than specialized skills within a single company. If progression with the company is largely based on knowing people in the company, it will add less flexibility than if it is based on the breadth of knowledge and skills. Jobs that allow employees to take part in different businesses or different subsidiaries at the same time will result in higher flexibility. This is also a good policy for the employer as the company can create more flexible employees, who can adapt to new jobs as they become available. If the company were to venture into new products or services in the future, having a larger number of employees with adaptable and general skills is more advantageous.

Individuals, companies, and countries try to avoid or reduce uncertainty, as this helps them plan and execute easier. Managers of companies prefer no or low uncertainty because they can manage their businesses with less effort. In an integrated world economy, uncertainty emanates from many different angles. Shocks in any part of the world quickly spread through the entire system, affecting the ability of companies to manage the demand for their products and services, the supply of raw materials, energy, and other critical inputs as well as the availability of financial and real (land, equipment, people) resources to conduct its business, regardless of its location, size, and scope. The world economy is a nonlinear system, akin to the weather, in which small disturbances or events can get amplified and produce catastrophic outcomes in different locations. In spite of an exponential growth in computing power, we are still unable to precisely predict weather more than a few days in advance, let alone speculate on long-term climatic changes. Although we are incrementally successful in forecasting catastrophic events such as hurricanes, these are extremely short-term phenomena and localized to a small region. The world economic system is very similar. Although it is segmented into countries, trade blocks and alliances, these are not self-sufficient units. We may be able to forecast localized demand and supply disruptions in short periods of time and associated price effects, but, just as in hurricane forecasts, they are short-term and localized. Countries, trade blocks, and alliances interact with each other through trade in financial assets (currencies, investments) and real assets (products, services, technologies, companies). Events in one country very often spread to others and get amplified. Uncertainty, thus, is on the rise for all businesses and individuals as they are continuously exposed to events anywhere in the world.

Just as the oceans and the atmosphere critically interact in the formation and destruction of unpredictable weather systems, real and financial assets are getting intricately connected in the creation and elimination of economic shocks worldwide. Till now, financial and real assets have been kept separate and managed by different organizations and people. Financial instruments include currencies, stocks, bonds, options, forward/futures contracts, insurance, and so on, while real assets include everything else such as land, equipment, technologies, intellectual property, skills, grains, oil, metal, and others. Those who own real assets may buy or sell financial assets to manage the risk inherent in them and to manage uncertainty. Those who operate in the financial world may use financial instruments to speculate or to take positions in real assets they do not own. It is this interaction between speculators and hedgers (risk managers) that drives the economic system and they emanate from both the financial and real economy simultaneously. From a systemic perspective, it is futile to effect a change in the financial world without considering corresponding effects in the real economy and vice versa. In the recent financial meltdown, many incorrectly believed that the shocks to the financial system could be localized but experience tells us that such shocks quickly propagate into the real economy. Similarly, it is impossible to contain a real economy shock, such as the rapid price decline of real estate prices, without affecting the financial system. The real and financial economies are two sides of the same coin and cannot be treated separately and this has implications for the level of uncertainty faced by companies.

A meteoric increase in financial innovation in the last two decades has allowed a closer integration of the world economy. Financial instruments that derive value from the underlying real asset (such as the mortgage backed security—MBS) provide a mechanism for the disaggregation, repackaging, and transaction of risk. There are many nonlinear feedback loops between the financial instrument and the real asset. For example, the availability of money that can be invested in the MBS instruments will drive the price of not only the security but also the underlying real estate. This is because such investments make the purchase and sale of real estate easier and the pool of buyers of real estate larger. In the recent financial crisis, investment banks buoyed by the leverage granted to them by the regulators, transacted in financial instruments (such as MBS), which in turn created a demand for transactions in real estate. This demand was met by both speculators acquiring real estate on leverage as well as a general lowering of credit standards needed to acquire real estate. As the pool of buyers grew, it drove up the prices of the real estate, which in turn affected the value of the financial security. As the reader can imagine, if played in reverse, this has disastrous consequences with both the real estate and the financial instrument based on it, falling in value in a downward spiral.

The physical constraints associated with the real asset such as its location and production characteristics (for example, the timing of harvest), do not preclude the purchase and sale of the economic value associated with it anytime and anywhere as this can be done through financial instruments. This complete linking of financial and real assets has also allowed the entire world to take part in speculation and hedging. This allows transactions at a scale and scope hitherto unimaginable, but in the process creates a much higher level of uncertainty within the system. Since financial instruments (unlike real assets) can flow without constraints across country and trade block boundaries, they are able to amplify small disruptions in some parts of the economic system and transmit them across the entire world quickly.

The feedback effects from the financial shocks on the real asset, which in turn affect the financial derivatives based on them, can create a spiral in value loss or a bubble in value increase. The scale of these unstable phenomena (similar to a hurricane in weather systems) is large in the current interconnected world economy and is expected to continue to grow as more countries join free trade treaties and open their financial systems to worldwide participation. Based on the recent financial crisis, many have come to a premature and incorrect conclusion that such integration is a "bad," and have begun to suggest that trade in real assets be localized and innovation in financial assets be stopped. This is based on mistaken impressions of the causes of the financial distress many countries recently experienced. It is neither the integration of economies worldwide nor the innovation in financial instruments leading to a tighter coupling of financial and real assets that caused the meltdown. Rather, it was due to the special favors sought and gained by a handful of companies from regulators (that resulted in high leverage in financial instruments) and the actions taken by the corrupt and incompetent executives in these companies. This was aided by the transactions in the real world where a general lowering of credit standards and mindless speculation by some led to high demand. If we are able to quell the world's anger at the incompetence and greed of the precious few who caused the problems and not to do any damage to the well-established principles of worldwide integration, trade, and interlinking of real and financial assets, we can soon return to an era of growth.

In the modern world, the scientific and engineering process of analysis dominates. One important characteristic of this process is breaking a system down into its component parts and analyzing each part in detail in an effort to understand the whole system. For example, in engineering, a method called finite element analysis (FEA) is employed to analyze the mechanical behavior of complex systems. In this method, a machine part or building structure is modeled in a computer by systematically building it from the components—very much like solving a puzzle from a large number of pieces that define it. Each small piece (an element) assumes certain predetermined characteristics such as how it will act or how it will move past others, in the presence of an external force. The behavior of each piece is then accumulated into predicting how the whole system will behave under certain conditions. However, this method has not worked well in highly nonlinear systems such as biology, economics, and business, where the properties of the individual components cannot be aggregated systematically because of the nonlinear interactions among them. Thus, the attempted applications of the scientific process and engineering methods in business to predict system behavior have been a failure (see Figure 1.2).

Many accept that organic system behavior is not easily replicated by unchanging laws that govern physical systems. However, this is still the norm in many industries today. In the financial services industry, the same approach to risk management has led many companies to catastrophic failure

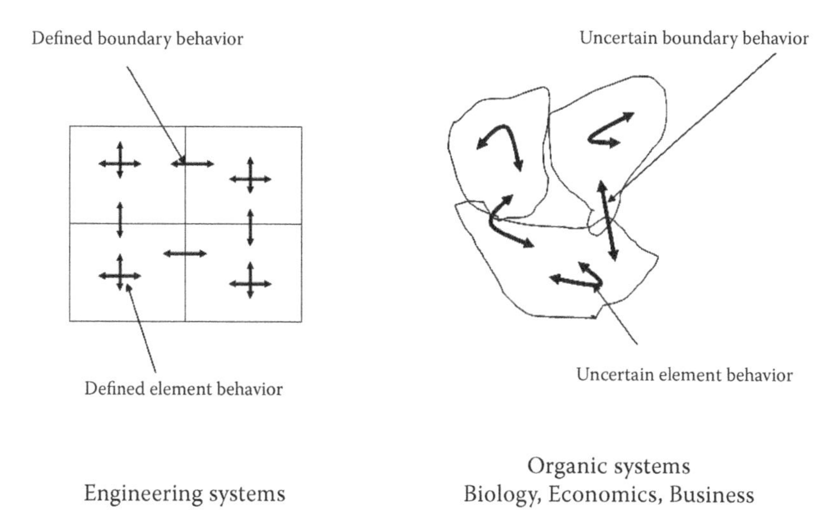

FIGURE 1.2
Engineering and organic systems and interactions.

recently. Managing risk in well-defined components, instruments, portfolios, and departments and then aggregating them, just as in an engineering system design, has spelled trouble for investment banks and private equity firms. Companies and individuals have to get away from the obsession of predicting the future precisely and put more effort into designing flexibility to manage future uncertainty. Predicting the future behavior of nonlinear and interacting systems based on initial conditions and postulated future events is a waste of time. More importantly, the confidence that comes from such analyses and processes may lead companies to engage in activities that will eventually lead them to complete failure.

While accepting uncertainty as important, some have attempted to differentiate between what can be reasonably predicted and what is unknown. However, such a thought process will reinforce the engineering view of systems. By segmenting what is known and what is unknown, apparent precision may be engineered into the known attributes. Many companies failed in the recent crisis not because they were hit by the complete unknowns but because of their inability to include uncertainty in the known attributes. It is not just the presence of "black swans" that is problematic, rather the assumption that the "white swans" walk in a straight line. When investment banks stuffed their balance sheets with the securities they knew well, they did not allow for a market-based valuation that would have captured the inherent uncertainty in them. They also, somehow, forgot the nonlinear interactions between different accounts in the balance sheet and the possibility that such interactions can be self-reinforcing.

When uncertainty is present, hypothesis testing is sometimes employed to statistically accept or reject a postulated aspect or characteristic of a system.

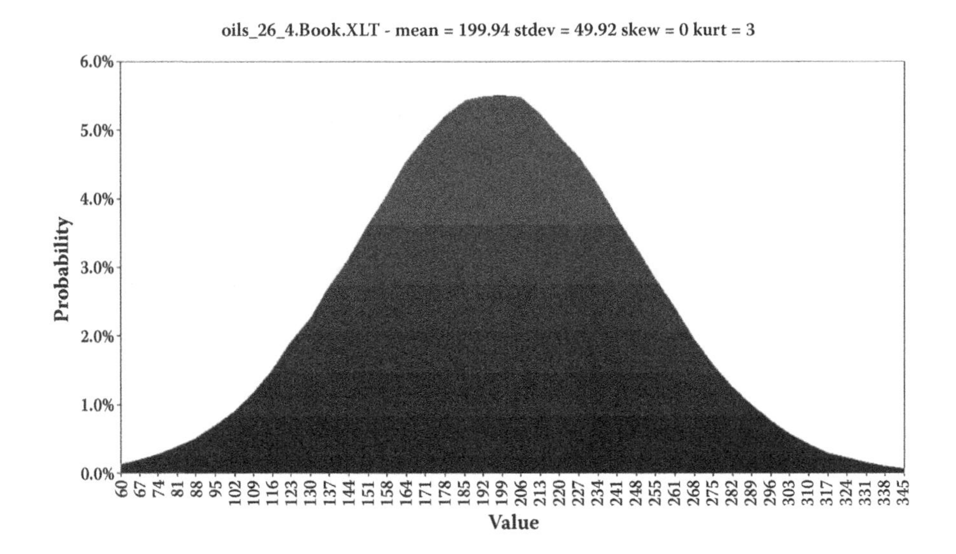

FIGURE 1.3
Normal probability distribution.

The hypothesis testing is based on the assumption of normality. This means that to accept or reject a hypothesis, one calculates the probability of the future occurrence of an event assuming that it is draw form a normal distribution. Readers may be familiar with the classical bell curve (normal distribution) (see Figure 1.3) with perfect symmetry on both sides and defined probabilities of certain events happening. The normal probability distribution is used to make decisions in many industries including life sciences, financial services, and technology. One of the reasons is that it has good mathematical properties that allow clean decision making, whether such decisions are good or not. More importantly, companies and regulators have been using it for a long period of time and deviating from these traditions is not comfortable for many.

This singular idea that the probability of a future event can be predicted based on a normal distribution has led to underperformance and failure in many industries such as life sciences and financial services. It may also be the case that hypothesis testing based on this assumption forces the discarding of information in the tails of an observation. The idea is that if you find the outlying observations (in either end of the bell curve) are less than some set percentage of total observations, decisions can be made assuming that such outliers will not happen. This method of decision making has two important drawbacks.

1. If the effect of the outlier is significantly higher than the others, it does not matter if the probability of occurrence is low. If the outlier happens, the failure is catastrophic.

2. It is the information content in the outlier that is most important for decisions. The fact that a large number of observations are seen in the middle may not have any relevance for decisions. The peak in the middle lulls decision makers into a false sense of comfort and routinely leads them to the wrong decisions.

Significant effort is expended to model complex systems using mathematical equations with precise answers and to use statistical measurements to drive toward decisions when uncertainty is seen. Both have led companies and individuals to wrong decisions routinely. The recent catastrophic failures in financial institutions due to segmented and conventional risk management, declining R&D productivity in information-rich industries where outlier information is routinely discarded, and declining rate of fundamental innovation due to education systems that force students to the middle while simultaneously attempting to increase the average and reduce standard deviations, are all examples of the perils of a conventional mind-set in analysis and decision making.

As the world gets smaller and the differences between real and financial assets become increasingly blurry, we have to accept that uncertainty will increase. Free flow of information and the ability to quickly react to it through financial instruments will amplify small shocks in the economy. Organizations should take existing and increasing uncertainty in all aspects of business as a given. Successful organizations will create structure, systems, and strategies to cope with uncertainty rather than focusing on any attempts to quell it. It is impossible to avoid or ameliorate uncertainty in the modern world, just as we have been unable to architect stability in weather systems. Uncertainty is not "bad," but rather it is an opportunity for those organizations that exhibit flexibility to create value by managing it.

As an example of managing uncertainty, consider the Aluminum Can Company, a U.S. company that produces aluminum cans and sells them to food packaging and distributing companies. This company is subjected to many uncertainties such as the cost and availability of raw materials (aluminum), cost of electricity and fuel used in the production processes, and the demand and price for the finished good (aluminum cans). Some of these uncertainties are driven by macroeconomic factors. For example, the cost of aluminum as well as the price of electricity and fuel may be driven by supply and demand and inflation that may affect overall price levels. The supply and demand are also driven by the business cycle. The company is likely a "price taker" in these markets for raw materials and it cannot influence prices much by itself since it is a small player in the global market for aluminum. The United States is the largest producer of aluminum with over $30 billion in use and exports, with worldwide production doubling that, but only about 10% of this production actually goes into aluminum cans.

The prices of commodities such as aluminum also exhibit a tendency to "mean revert" to a long run average price. When prices are higher than typical, the miners work overtime and marginal mines are put into operation, increasing the supply of bauxite (hydrated aluminum oxide), the aluminum ore. Aided by the higher price of aluminum, the processing plants that make aluminum from bauxite also increase output. These plants first convert bauxite into alumina and then through an electrolysis process eventually to usable aluminum. New plants may also be constructed by companies trying to take advantage of the price rise. The supply of aluminum thus increases. This will curb the price rise and will eventually drive the prices down as supply exceeds demand. As the prices fall, many of the marginal mining and production operations shut down, reducing supply. This will arrest the fall of aluminum prices as supply decreases. This will eventually increase aluminum prices. This tendency to pull the prices lower when it is high and push it higher when it is low is called "mean reversion." Although we know mean reversion exists in commodity prices, this alone does not allow us to precisely predict future prices of aluminum. The demand and supply dynamics depend on many factors and so the future prices of the commodity will remain uncertain. Since aluminum is an input into many products, including airplanes, automotive, lighting, electronics, and music, the demand patterns in many industries affect the overall price of aluminum. Note that the prices of the financial instruments that can be used to speculate on the price of aluminum (such as futures and forward contracts) reflect the overall expectation of demand and supply of aluminum today and in the future. The forward curve of aluminum prices is an expectation of future prices, given all currently known information. However, as the future unwinds, events will happen and new information will become available and the actual prices will be different from what is seen in the current expectations about the future.

On the demand side, since its own end product (aluminum cans) is an input into the production processes of its customers, the demand is related to how well its customers' products are selling. For example, the company's customers may include canned food makers who supply to grocery stores. So, the demand for aluminum cans ultimately is a function of the demand for canned food. If the buyers of the company's products are largely domestic, the state of the domestic economy and customer taste may drive such demand. Americans use over 100 billion cans per year, with worldwide demand approximately twice that. Such demand can be countercyclical to the macroeconomy, meaning the demand may increase in recessions when people stay at home more and cook rather than eat out. The demand for aluminum cans is also affected by consumer taste. U.S. consumers currently prefer aluminum cans to glass or bottles and tend to recycle them more. But this can change in the future. Consumer tastes are fickle and the younger generation may start to prefer glass, for example, and if so, this has a negative effect on the demand for aluminum cans. Equally possible are the advent of new packaging materials and innovation in food and packaging that may even make hard packaging obsolete. This single product company, thus, has significant uncertainty in all aspects of its business, most of which is driven by factors outside its own control.

Most companies are much more complex than the one described here and the combined uncertainties faced by them are many orders higher. The managers of the Aluminum Can Company have to manage all the uncertainties they face well to create economic value for their shareholders. If such uncertainties did not exist, companies do not need managers. The owners of the company could simply invest in machines that produce the finished goods and enter into long-term contracts with suppliers, buyers, and employees. They can then sit back and receive a predictable stream of profits. Since this is not the case with most businesses, the owners (shareholders) put managers in place to (primarily) manage uncertainty. Those who have the skills to manage uncertainty will be rewarded. However, it does not work that way always. Managers of the company may be more interested in managing uncertainties in their own careers than the uncertainties faced by the company.

How does one manage uncertainty? Uncertainty is managed by either removing the cause for it or by introducing flexibility to manage through it. For example, the uncertainty in the price of electricity can be removed by entering into a fixed-price forward contract for it. The company can essentially hedge the price uncertainty of electricity so that it does not have to manage it. But removing uncertainty is a costly activity for two reasons. First, to hedge the price of electricity, the company has to enter into a contract with somebody else (the counterparty) who is willing to do so. The counterparty will be willing to enter into such a contract only if it is paid an adequate compensation for taking such a price risk. If the price of electricity increases in the future, the counterparty who guaranteed a fixed price will lose. However, if the electricity prices drop in the future, the counterparty makes money from the fixed-price contract. In a market where the uncertainty in electricity prices can be mathematically determined (one can draw out the probability distribution of future electricity prices based on demand/supply data and past trends), such contracts will be priced efficiently (neither party can take advantage of the other). Since electricity cannot be stored efficiently, the price of electricity shows sudden jumps (when demand is higher than supply). Electricity prices can also show seasonal variations depending on location and whether it is used for heating or cooling households and factories. The seller of the fixed-price contract has to consider such price jumps (that may last only hours or minutes), seasonal as well as long-term trends in prices. The seller of the hedging contract will price the contract in such a way that the price risk taken is compensated appropriately. In other words, the company will have to pay a price for the hedging contract that is roughly equal to the economic value of such a contract given the future uncertainty in electricity prices. If the contract is priced lower, many companies will be willing to buy such hedging contracts or not too many speculators will be available to sell such contracts, driving the price of such contracts up. If it is priced higher, the reverse will happen.

Second, by hedging the price of electricity (and thus removing the price uncertainty), the company many have increased the risk of its own business. Imagine a situation in which the demand of aluminum cans and the price of electricity are correlated with the U.S. economy. This means that if the U.S. economy is doing well, both the price of electricity and the demand for aluminum cans (and thus the price per unit that can be demanded) will be higher than if the economy were in recession. If the company enters into a hedging contract for electricity (and thus fixing the price of electricity), it will be better off in situations where the economy does well (cost is kept low and revenues are high). However, if the economy enters a recession, the company will be much worse off as its costs will remain high (thanks to the fixed-price electricity contract) and its revenue will decline (because of lower demand for aluminum cans).

Similarly, consider an airline that has entered into a forward fixed-price contract for fuel. The company will do well if the economy does well, helped by its hedging program keeping the price of fuel low while demand for flying allows it to increase the price per available mile flown. However, if the economy stalls, the company will be hurt worse than its competitors (who do not have such hedging programs). In this case, the company will be obligated to pay the high fuel prices because of the forward contract in place when the oil prices decline, helped by the slowing economy. So, removing uncertainty through financial hedging as described above is costly and it can increase the total risk in the business.

Companies can remove uncertainty altogether by hedging inputs, conversion costs, and outputs. For example, consider a paper company that has entered into a contract to sell a fixed quantity of designated paper at a fixed price for the next several years. During the same time, it also entered into fixed quantity and price contracts with the producers of pulp and other inputs into the production processes. The company already has machines in place and a fixed-price maintenance contract with another company. It has also struck a deal with the union for a long-term contract for labor that fixes wages and benefits. Finally, it also entered into fixed-price contracts for fuel and electricity that will be used in the production processes. This company, thus, has removed all uncertainties that it currently sees. Is this a good way to manage a business? There may be two problems here. In the short run, the company has turned itself into a financial instrument. Since the revenue and costs are fixed, its cash flows are fixed and it will behave more like a bond instrument than a company. Such a company may not need managers if the contracts are static and defined. If the company has investors, the stock price of the company can be determined by simple discounted cash flow analysis. In the long run, as it has eliminated uncertainty in everything, it may also have eliminated its ability to change or adapt. If a new substitute for paper was discovered or a new technology was invented to incorporate a special kind of paper in computer chips, the company will be unable to adapt to it. In the first case, when the contracts run out, it will simply close. In the second case, it will be left behind in profitability as its competitors take advantage of the newly created demand. So, removing uncertainty is not necessarily a good way to manage a business.

Introducing flexibility is another way to manage uncertainty. What is flexibility? Flexibility is an attribute that allows companies to manage through uncertainty. For example, if the Aluminum Can Company had a labor contract that allowed it to increase and decrease employment without penalty, this would introduce flexibility in managing the company's labor costs. As the demand for its products increase, it can increase employment (and trade off overtime as necessary). On the other hand, in tough times, it can decrease its labor cost by reducing employment. Such labor flexibility allows the managers to make labor decisions as uncertain demand is revealed. Readers may be familiar with the low labor flexibility that exists in countries like France where managers face restrictions from the government and regulators in dismissing employees. Such government-induced inflexibility increases the total uncertainty in the company's profitability. Managers of companies in such a country will delay hiring decisions as such decisions cannot be easily reversed. Since hiring is an option (a right but not an obligation) for the company, it will not exercise it prematurely when uncertainty is high for its products and thus for its demand for labor. This is reflected in the persistently high unemployment rates in such countries. Note that rigid labor laws are similar to financial hedges that remove uncertainty in certain aspects of the operations (such as total number of employees) but in the process also remove flexibility, increasing the overall risk in the company. If regulations related to dismissal of employees when demand is low are removed, companies can operate with a higher level of flexibility and better manage uncertainty. Similarly, introducing an option to use multiple types of fuels—say oil and natural gas in its production processes—will be a way to increase flexibility. In this case, the company can select a lower-cost fuel in the future, after having observed the prices for both. If oil prices are high, it may opt to run its plants using natural gas and vice versa. Not having hedging programs in place also introduces a natural flexibility into managing uncertainty. The company's fuel purchases will scale up and down along with the demand for its products.

Although introducing flexibility into labor contracts and adding a dual fuel option in production processes may help the company to better manage short-term uncertainty, it also has to think about flexibility more strategically. If the company's objective is to succeed in the long run, it has to define its objectives clearly. For example, the Aluminum Can Company may define itself as a leader in the application of aluminum (cans being one of them), in the packaging of perishable goods or in other ways. If the focus of the company is aluminum, it has to constantly invent better products around aluminum. If its focus is packaging, it has to find better ways to package goods, aluminum cans being one such method. In either case, it has to have sufficient flexibility to innovate. A focus on efficient production of aluminum cans, albeit being good for tactical profits, may not help the company later as consumer trends change or other innovations make aluminum cans obsolete. A focus on reduction of uncertainty in status quo may make the company unaware of what is happening in its markets and the world at large. The company has to expose itself to uncertainty but add flexibility to manage it both in the short term (price and quantity uncertainties) and in the long term (consumer, product, and technology uncertainties). By creating networks that include its consumers and suppliers, the company can be in constant touch with changes both in end products as well as in materials. By designing the company to be part of a flexible network, it can obtain new information faster and adapt to it continuously.

Uncertainty and flexibility are relevant not only for companies but also for countries. Policy makers have to consider flexibility as an important strategic lever in making long-term policies. Since macro-uncertainty exists in many dimensions, policies that are designed to deal with precisely predicted future forecasts will remove flexibility and reduce overall value to society. For example, an energy policy that is based on precise forecasts of energy consumption per capita, population growth, and supply of fuels may provide incorrect incentives for future production. For example, subsidies for ethanol were introduced in the United States when oil prices were high and these induced investments in ethanol production facilities in the United States. Ethanol production may be competitive if oil prices are above certain price levels but not otherwise. Also energy use is driven by the use of prevalent technology and future (and currently unknown) technologies. For example, if technologies are invented to reduce energy consumption in transportation, lighting, heating, and data storage, it can drive down the current per capita consumption of energy. If the past policies have resulted in an overinvestment into the ethanol infrastructure, it will render the overall system rigid, unable to switch to a more efficient means of producing, transporting, and using energy. In general, policies that provide incentives for high and rigid capital investments (based on current expectations of demand and supply) will reduce flexibility for future energy management. Policies that induce innovation of transformable and switchable production methods will increase flexibility and allow the system to smoothly transition to future demand and types of use. They will be able to introduce flexibility into energy management to deal with uncertainty better and such technologies will be able to cope with whatever future outcomes are.

Active monetary and fiscal policies that are based on precise expectations of the future reduce overall flexibility of the macroeconomic system to adjust to uncertainty. Consider the setting of interest rates by monetary authorities. Setting a specific interest rate and changing it up or down at predetermined intervals so as to keep the economy balanced on full employment and low inflation is a bit like trying to balance a soccer ball on a cue tip. By setting a high interest rate when inflation is high or a low interest rate when unemployment is high, for a fixed period of time, the monetary authorities constrain the macroeconomy from continuously adjusting to changes and known information. Instead, in a passive monetary policy regime, a continuously adjusting (and more flexible) economy can be created by letting the market mechanism set the rates. It is preposterous to assume that a few bureaucrats can determine optimal interest rates better than millions of market participants. It is the ability to change quickly (i.e., flexibility) that is more importantly than the infinite wisdom doled out by the policy makers in an attempt to drive the $10 trillion U.S. economy.

Similarly, fiscal policies that shift decision making to the government in capital allocation across sectors likely will reduce overall flexibility for a country. Since fiscal policies are determined by a small group of people, it is unlikely that they will have full information regarding optimal allocation. Government bodies are also well known for their proclivity to allocate capital to pet projects, regardless of merit. They are also slow in reacting to changes in the economy. A market-based system that is fast in understanding changes and optimally allocating capital (driven by incentives) affords higher flexibility. The amount of spending by the government in the economy is a proxy for fiscal rigidity and slow market adjustments to new information. The less government spending is, the more flexible the economy is likely to be.

Uncertainty is an important characteristic of today's environment and flexibility is an important aspect for managing it. They are related and those who understand them and manage them well are likely to increase shareholder and societal value. Managing uncertainty and flexibility is a complex exercise. Most common notions of management fail to do this. One of the reasons for this is the prolongation of ideas that have been with us for a long time. In this book, I will dissect organizations into their components, identify types of uncertainties faced by them, and describe how organizational features can be designed or altered to better manage uncertainty. We will challenge status quo ideas in all aspects of organizations and decisions.

Throughout this book, I use *flexibility* as the primary vehicle to improve the functioning and value of an organization. Wherever appropriate, I will attempt to describe the status quo from a historical perspective. This allows us to identify why certain organizational aspects exist today even though they may have been rendered inefficient or irrelevant due to changes in the world, business, and human aspirations. I will investigate how flexibility can be designed in many aspects of an organization such as structure, system, and strategy.

2

The Evolution of Organizations and the Environment

The Industrial Revolution, beginning in the late eighteenth century, created significant changes in technology, organization, and the society as a whole. Steam power, supported by fuels such as coal, influenced the invention and implementation of new technologies. This ushered in an era of a production economy in which a single powerful idea resulted in gigantic industries and huge enterprises that thrived on scale. In the scale economy, efficiency was very important, and managers had to focus on getting the conveyor belt moving in the most optimal fashion. Strict time records were kept, and wages were only functions of duration of work or production. The contract between the employee and owner was simple—wages for time worked or goods produced. Measurement was reasonably straightforward—recorded time on punched cards or paper or a count of produced units of goods. This was sufficient to execute the wages part of the contract. Work was characterized as a commodity product—a substitute could be found with relative ease. This was an era of determinism. Many of the skills of the scale economy manager—a strict adherence to rules, error-free record keeping, and the handling of money—were different from the skills of a production worker. These skills made the manager less of a commodity. This is the beginning of the skills-based class structure in large enterprises. When the owner noticed such a difference (as did the manager), the contract between them became a bit more complicated than the worker–owner contract.

One of the differences in the manager's job, however, was the criticality of the tasks performed. The downside risk of underperformance of the manager was higher (from the owner's standpoint). The owner thus was willing to take insurance against such downside risk through a positive reinforcement of an attendance bonus or a negative reinforcement, such as a threat of job loss for absenteeism. As complexity of production (such as skills specialization in workers) increased, the manager's role began to take a higher dominance in the overall system. The manager (because of intimate knowledge of workers) held information such as individual worker productivity and the proclivity toward absenteeism that allowed the owner to optimize production. This upside potential further drove the owner to enter into complex contracts with the managers that included a production-based bonus and profit sharing. Such contracts began to show complex embedded options for both parties to manage uncertainty. All these trends increased the managers' power in the enterprise. Quality of management was then measured by the manager's ability to produce more work (according to relatively stable specifications and simple instructions).

In the mid-twentieth century, as the world wars ended, economic wealth increased, industries matured, and consumers became more knowledgeable, a new competitive arena, differentiation, became feasible. In this environment, value became a function of features and utility in addition to weight and numbers. The power of a single idea diminished, and integration of a sequence of ideas began to command higher competitive leverage. These ideas were extensions of an existing dominant idea or the bundling of one or more existing ideas. Since the owner's value was now a function of production scale and scope of products, the owner–manager contract needed to be constructed to consider both aspects. Overall profitability of the owner was dependent not only on aggregate production and efficiency but also on the rate of improvements made to the original idea. The value of the improvement was a function of both timing (how quick) and relevance (how profitable).

The initial reaction of the owner to the changing environment was the introduction of higher specialization of workers. Production and design departments were created and kept apart to ensure clean lines of accountability. Manager specialization was increasing as well. Production and design managers with different skills were sought even though the fundamental role of either manager was about the same, namely, higher production. The production worker and manager contracts were also similar in the new environment. For the design worker and manager, the measurement complexity was higher. Since both the timing and the relevance of the specification change were important for owner profitability, incentives had to be designed to optimize both. These changes in production specifications led to higher inefficiencies in production, and the scale advantages, on which the whole enterprise was based, were slowly disappearing for the owner and for the production manager as well. If the worker had a strict quantity contract, his or her compensation declined (as a changed specification took more time to produce and the brain needed to be retrained for a different set of routine activities). This incremental disutility was compensated by the appearance of variety for the production worker for the first time. The possibility of replacing a set of routine activities by another similar set of routine activities at regular intervals added utility to at least a cohort of production workers.

For the production manager, this posed new challenges. Measurement of worker productivity became more complex, and calculation of compensation from quantity and time worked were not enough. Worker productivity did not follow a linear relationship with experience (some were more adept at changing quickly than producing more of the same specification). With these challenges came new information to the production manager that could be used to increase profitably for the owner. For the first time, both production worker and manager contracts changed to take into account the ability to change quickly into producing an incrementally different specification drawn by the design department. Technological change during this time also lent a major blow to the scale-based enterprise. The idea of higher efficiency through increased scale was challenged by the ability to achieve rapid customization through technology.

By the late twentieth century, there was more tension looming on the horizon for enterprises: a tension between the design and production departments. Complexity in contracts led to misaligned incentives between these two groups. There was no incentive for the design department to ensure that changing specification did not lead to production difficulties and ensuing erosion of profits for the owner. Forward thinking owners attempted to combine the departments or align incentives more closely to overall profits. Such owners quickly realized that design and production were not independent activities, and that production was likely to be the best place to extract design ideas. Leveraging the worker knowledge (now fashionably called worker empowerment) helped many companies increase productivity. In all this chaos, managers gained more power as the owners lacked information to reach conclusions on optimal structures.

Unfortunately, the managers' utility curve had competing priorities. One of these is the desire to have a large group of people working for the manager. This was partly the result of incentives in the contracts that the owner imposed on the manager. The power base was a function of the number of workers reporting to the manager as the owners' expectation of the managers' knowledge was correlated with this number. This scale mentality led the owner to make incorrect decisions, and the owner was unable to maximize value. The spread of ownership through mature equity markets and the efficiency of information flow finally provided a new way out of this manager–owner stalemate. One outcome was the removal of managers whose utility curves were in conflict with overall firm value. To the surprise of owners, the scale and scope economy had created a huge growth in the number of managers, and the removal of a substantial part of this management layer actually increased firm profitability. This manager downsizing was also accompanied by the breaking up of large enterprises into smaller entrepreneurial units with higher alignment of incentives among workers, managers, and owners. This downsizing of large companies also resulted in simplified organizations. The reduction in complexity enhanced the ability to manage the enterprise under uncertainty and this brought huge incremental value to the overall economy.

During this time, the business environment was experiencing the third major change, one that is driven by innovation. Industry finally arrived at a point in time that recognized that the brain is in fact more powerful than brawn. Scale from the production of powerful single ideas and the incremental extensions of such ideas through scope gave way to the fundamental need for innovation. This was the first real break from the ideas that have been with us for so long from the Industrial Revolution. The innovation economy is fundamentally different from the previous situations. In this environment, there was little role for the production worker or manager who performed essentially routine activities. Technology was available for automation of the routine, and leveraging the skills of the workers into innovation was the most important competitive lever. The transition into the innovation economy provided opportunities for companies to configure their value chains, taking each other's strengths and weaknesses into account.

In the innovation economy, shortening product life cycles, diminishing economics of product extensions, and accelerating wealth of demanding and discriminating consumers all point to a very different way of doing business. The competitive advantage is solely dependent on how fast and how profitably new ideas are generated in the new environment. It is not the quantity of routine production but the quantity of essentially nonroutine innovation that is important. In this new era, driven primarily by the rate of innovation, the traditional management processes and techniques need to be changed. The contract between owners and managers needs to be redesigned, and the incentive system needs to be rethought.

One important idea in this transformation process is flexibility. In the scale and scope economy driven by efficiency, we strive to reduce flexibility as it hinders orderly production. However, if innovation is the driving factor, flexibility has to take the front seat. Flexibility has to be used as a fundamental guiding principle—in organizational design, incentive and contract designs, human resources, infrastructure, planning and scheduling, and capital structure—in every aspect of business. What does it mean in practice? Business thinkers have been using the term *organizational agility* to describe the ability of the organization to quickly adapt to changing environments, competition, and technology convergence. Flexibility is an organizing principle to enhance agility. Flexibility allows an organization to both identify trends early and adapt successfully. In general, as the size of an organization increases, flexibility decreases. One way to stay flexible is to design the organization in smaller entrepreneurial units that are tied together not by rules but by culture and competence.

Human resources are treated as an afterthought in many companies, and the term implies people management and not value enhancement. In the innovation economy, human resources assume critical importance and its treatment as a commodity in the scale and scope economy needs to be abandoned. Innovation is a result of both flashes of brilliance of individual creativity and systematic problem solving in groups of varied competencies, skills, and expertise. Hiring based on set specifications and offices that resemble jail cells no longer work. Introducing higher levels of flexibility in human resources with a variety of skills and expertise and designing a work style that draws in more unconventional participants (who may not adhere to the standard 9–5 rules) are important for the development of a workforce capable of innovating at an accelerating rate. Companies also have to realize that diversity is an important ingredient for flexibility. Human resource policies that encourage similarity in thinking, education, location, and physical characteristics will reduce flexibility and make the company vulnerable to failure in uncertain times. Just as a cloned strain of corn, optimized for kernel production, will increase unit productivity, it will also be destroyed by a disease that may not affect variant strains.

Incentive and contract designs are also increasingly important in the innovation economy. For example, configuration of the value chain of the company, including its suppliers and buyers, will need to internalize a system-wide flexibility. Redundancy in critical suppliers, ability to increase and decrease purchases as new information is revealed, ability to reroute and reconfigure in the event of major disruptions, and the ability to switch products, locations, and processes are all part of flexibility. Such system-wide flexibility needs to be reflected in better contract designs between partners that incorporate flexibility systematically and price them efficiently.

Design of incentives has shown little innovation in recent times as most companies follow set principles such as stock options and top-down pyramidal compensation schemes inside the company and rigid contracts with suppliers and buyers outside. None of these have shown to be particularly effective in increasing productivity and shareholder value. One of the primary reasons is that these instruments are designed more for the scale and scope economy, and they generally fail in the innovation era. New instruments and methods need to be designed that will enhance innovation by the creation, nourishment, and optimal exercise of real options. Incentives should encourage better management of uncertainty rather than attempts at eliminating uncertainty. Decision makers have to understand that uncertainty is a source of value if managed with flexibility, and conventional notions of minimizing risk at any cost will lead them down the path to mediocrity. Optimal risk taking should be encouraged and failure accepted, not shunned.

Corporate finance also has been stagnant in recent years in many aspects of capital structure, project financing, and portfolio management. The general tendency has been to apply accepted theory and assume that the capital providers also follow such ideas. For example, a significant amount of time and effort of senior decision makers is wasted in the "management" and "communication" of quarterly earnings. Although empirical evidence has been strong that enterprise value is a function of the future prospects of the company and the ability of its management to create, nourish, and optimally exercise options, many still act as if value is driven by a blind discounting of earnings of this quarter and projected earnings of future quarters.

Companies have to depart from the idea that real asset decisions can be fully disconnected from the financing decisions and create a holistic portfolio management process that includes both. Management of the company in silos—where research and development (R&D) decisions are made disconnected from financing decisions and capital structure decisions are made in the absence of a good understanding of the risk of the real asset portfolio—is fraught with danger. It is clear that the "local optimization" attempted by large companies—in providing incentives for R&D to increase throughput in product extensions, marketing to increase volume, finance to meet the numbers—is by definition suboptimal for the owners of the firm.

The infrastructures of many large companies today reflect the rigid engineering designs that have been with us since the Industrial Revolution. Those who moved forward faster, shunning legacy ideas of production, have been able to use infrastructure planning as a competitive advantage in itself. Toyota's ascendancy in automotive manufacturing is a good example of early identification of the introduction of flexibility in manufacturing, including technology, location, and product design. Similarly, Hewlett-Packard's application of "postponement"—the ability to avoid full assembly of all components until the product is ready to be delivered—required a rethinking of manufacturing and logistic infrastructure across the world. Companies have to almost religiously apply options thinking in designing and managing infrastructure in the modern world, fraught with disruptions caused by natural and human-made disasters. Planning, scheduling, and management—in both manufacturing and service industries—seem to depend a lot on historical data. Again, we see remnants of past experiences from the scale and scope economy at work. Arrival of computing power at almost zero cost allowed installation of large collection bins of historical data in the belief that such data will always allow better planning and management. In the last decade, enterprise resource management (ERM) systems have proliferated, and we now have data piling up in data warehouses at an exponential rate. It is not clear, however, how such data are used in reaching decisions and whether more historical data actually enhance planning and management. Attempts at precise forecasting of the future based on historical data are futile. A major role of historical data is to provide information on uncertainty and risk and not on forecasting future outcomes precisely.

In an era driven primarily by innovation, companies have to focus on incorporating uncertainty and flexibility in all aspects of business. Flexible designs of organization, structure, systems, and strategy to manage unavoidable (and not to be avoided) uncertainty are necessary but not sufficient for success of companies in the innovation era. In addition, decision makers must make better decisions on how, when, and what to invest and how to maximize the value of their enterprises through holistic risk and portfolio management. They have to make decisions systematically based on market-based economic value and avoid the temptation to make ad hoc decisions. A holistic framework that considers uncertainty and flexibility is essential for better management of today's and tomorrow's companies.

3

Components of an Organization

In this chapter, I will analyze the major components of an organization—structure, system, and strategy. Structure is related to how a company is organized, system describes how it operates, and strategy entails how it influences the environment, both internally and externally. I consider these three—structure, system, and strategy—as the primary building blocks of an organization. I further break this schema down into their subcomponents as shown in Figure 3.1 and categorize them as tangible, semitangible, and intangible. This categorization is based not on physical aspects but on how the organization utilizes them.

Structure: Human Resources

By *structure*, I mean a much larger construct than organizational structure. It is related to all aspects of an organization such as human resources, information technology, and infrastructure. Human resources structure includes what an individual is equipped to do (skills and education), what environment is most effective for the individual (language, culture, and location), and what the individual would like to do (personality, objectives, and utility). Human resource organization along these dimensions will ensure that the happiness is high for the individual and the overall productivity is maximized for the organization. These characteristics are fundamental to

	Tangible	Semi-Tangible	Intangible
Structure	Infrastructure	Information	Human
System	Technology	Process	Content
Strategy	Internal	Boundary	External

FIGURE 3.1
Components of an organization.

the individual and thus to the organization that has assembled the human resources for a defined objective. They do not change unless the individual or the organization takes actions to change them, such as skills retraining and language classes. They are not related to the state of the economy or the prospects of the company's products and services. At this fundamental level, these dimensions represent "rigid" characteristics for the organization. For example, a company that has assembled a large number of English-speaking machinists who always worked in Detroit and love their jobs as machinists in Detroit has incorporated a level of rigidity (inflexibility) in the human resources structure. The human resources structure in this context is also not static. As the individuals' skills change—through education, experience, and understanding—the human resources structure of the company changes with it. Unlike contemporary structures where promotions and title changes drive structural changes, these types of changes are dynamic and more fundamental to the company's survival and success. For example, companies that consciously rotate employees to different jobs continuously will improve their skills and thus improve organizational structure flexibility. Similarly, a company that increases diversity in skills, type, location, gender, age, and other attributes of its human resources will enhance its human structure flexibility. The structure of the company does not end within it but has to include the network it is part of. Such a network may include suppliers and customers. Ultimately, the flexibility in the network has a great impact on the company's success and failure.

Structure: Information

Structure of information encompasses how the company collects, processes, and utilizes information. Information systems may create and use information structures but it does not start or end there. Information arrives at the company through a multiplicity of channels including but not limited to employees, the Internet, and other media. How the company collects such information is part of the information structure. Companies seldom have a unified way to collect disparate information and may focus on one or other aspect of information channels but not all. In many cases, companies may spend external resources to collect information from databases and consultants but it may already be resident in the company in richer detail, including in its own employees and experiences.

The explosion in computers and the reduction in prices for computing and data collection have led many companies to collect large amounts of data without consideration for the ultimate use for them. Collection of raw data without a coherent information structure increases noise and associated complexity. In many cases, information technology projects are initiated with a singular objective of collecting data, thus wasting valuable time and resources. Companies may have to not only define information in a holistic fashion (that encompasses the entire organization and not specific divisions, departments or employees) but also include all information channels. Data also should include the uncertainty surrounding them. For example, the company may have collected data on existing industry capacity for one of its products—say x units/month. This information may be lacking detail as capacity depends on employment, machine downtime, and other factors. As the industry capacity is uncertain, data related to it may be better defined by a probability distribution rather than a singular number.

How the collected information is processed by the company is equally important. Many companies typically internalize only static data. That is to say, spreadsheets have a familiar look to them as all numbers add up neatly—both in columns and in rows. Widely available spreadsheets and the use of them by managers may have introduced some rigidity into the way information is processed in companies. This is further exacerbated by the availability of charts and presentation programs targeted at demonstrating precise answers. Most available tools help process static information to make decisions or influence decision making. Consultants and analysts armed with spreadsheets can reduce flexibility in information processing, if they follow standard templates with constant data. Information has to be processed in such a fashion as to include uncertainty, interactions among information elements, and possible temporal effects. It spans space and time, is uncertain, and shows interactions. Typically, none of these attributes are considered in today's companies.

Also important is the storage modality of information. Cost of storage has declined fast in recent times. This coupled with the availability of enterprise resource planning (ERP) systems has led companies to store massive amounts of data in so called "data warehouses." Since data do not have a date of expiry, such data has been sitting in data warehouses for years—never used by the company in any fashion. However, information does have a date of expiry and if it cannot be used in real time, it will lose value quickly. Most information also changes quickly, making static storage of data a less valuable activity.

Structure: Infrastructure

Infrastructure includes physical (real) infrastructure and financial structure. If a company has constructed its own office buildings and manufacturing plants, it owns the physical infrastructure. The company can also lease physical infrastructure, which will be part of its financial structure as that will create lease obligations the company has to meet. In general, owning physical infrastructure will reduce flexibility as the company will not be able to increase or reduce its size dynamically as new information becomes available. How the company designs its physical infrastructure is also important. For example, a lighting system for the office building owned or leased by the company that uses sunlight (when available, perhaps channeled by fiber optics) as well as electricity has additional flexibility. Part of the infrastructure is work modality that includes how the company is structured legally, for example, as a limited liability partnership (LLP) with members or as a corporation with employees as well as how the company conducts its business. Some companies allow all participants to work from any location (location flexibility) and at any time (time flexibility). Others may have rigid schedules and work locations, thus reducing infrastructure flexibility.

How the different types of infrastructure owned by the company are configured is also important. For example, a company that owns many smaller manufacturing facilities in different locations may afford a higher level of flexibility rather than a single large facility in a single location. Ideas perpetuated from the Industrial Revolution such as scale-based cost advantages may force companies to design infrastructure so as to achieve small and tactical cost advantages. However, the long run scale-based designs may become a liability for companies as the business model changes. Flexibility can also be introduced through diversity of infrastructure. For example, a power generation company that owns some base-load plants that operates continuously and at lower generation cost per unit and peak-load plants that can be switched on and off (providing operating flexibility), albeit at higher generation cost per unit, has higher flexibility to manage uncertain future demand. If the company can use different fuels in its plants—say, oil and natural gas—it has flexibility in the selection of fuels to manage cost uncertainty better. If a company has the option to either produce internally or to buy from outside, based on price and cost, it has additional flexibility.

Flexibility is also a fundamental driving factor in risk management. Smaller manufacturing units in different locations allow the company to manage risk of loss of service and recover from catastrophic events. Some level of redundancy enhances the ability of the company to manage shocks. The infrastructure maintained by the company's suppliers and buyers and their relationship to the company's own infrastructure define its network flexibility. For example, a company that has suppliers in diverse locations adds flexibility in its real infrastructure (production and logistics) as well as its financial structure (such as currency exposure). Similarly, the pricing of its products (part of the financial structure) can be optimized by managing the entire supply chain network as a single system. In this case, network participants can be given various incentives to reduce cost or increase availability as demand changes.

In addition to product pricing, leases, currencies, and cash management, how the company is capitalized is an important aspect of flexibility. Start-up companies, for example, often are forced to seek venture capital. This may impose constraints on the company, perhaps due to standard templates followed by the capital provider. The venture capitalist's (VC) objectives may not be aligned with that of the company's and this leads to confusion, conflicts, and loss of flexibility. The VC's template may include a fixed expected return (such as 60%) and exit horizon requirement (such as five years) and they often adhere to these crude metrics without any appreciation for the dynamic aspects of the company and its environment. A company that has flexibility in other components of its structure such as human resources and information may render itself extremely rigid because of the financial structure, if the capital was sought from sources that have misaligned objectives. Start-up companies without sufficient capital, however, also suffer from a lack of flexibility. Seeking capital after certain events have happened is not a good strategy as the capital provider will seek a higher percentage of the company's value in a good scenario and refuse to fund the company in a bad scenario. Thus, it is important to capitalize the company at the right level and at the right time using capital providers whose objectives are well aligned with that of the company.

The debt in the capital structure of a company also reduces flexibility. Since debt payments are rigid, companies with higher debt have higher risk of bankruptcy in an uncertain environment. The financial flexibility loss created by high leverage (debt) taken by the investment banks was evident in the recent financial crisis. As the CEO of one of the investment banks famously remarked, "people are learning there are two sides to leverage." High leverage, although an easy way to multiply mediocre returns, may result in the complete failure of the company in bad future states (unless of course they are bailed out by the government). Similarly, cash and marketable securities provide higher flexibility for companies to take advantage of short-term opportunities presented by uncertainty. Companies that face higher levels of uncertainty in their business generally tend to have lower debt and higher proportion of cash in their financial structure. The cash allows the company to be tactically (short-term) flexible and lower debt provides long-term flexibility. Complex tax rules have played mayhem with the financial structure flexibility of large companies. For example, in the last decade companies were forced to keep cash abroad to lower taxes and avoid repatriation penalties. Such policies, if followed for a long period of time, can create significant segmentation in the balance sheet of the company, leading to a loss of financial flexibility. If the company does not have access to the cash quickly or using cash from abroad results in a tax penalty, presence of cash in its balance sheet does not result in any flexibility advantage. In an effort to satisfy short-term (and often meaningless) profitability, companies may pursue such policies, unaware of the loss of flexibility and associated value loss for the company.

Systems: Technology

All technology—information, manufacturing, logistics, energy, and other aspects of the business—come under the category of technology systems. As technology develops, there is a general convergence in many of these aspects. The technology systems employed by contemporary companies could be different in different departments, locations, and uses, thanks to disconnected decision making, disparate skills, and history of adoption. Differing technology systems can increase the complexity in the company and thus reduce flexibility. Inflexibility comes from the inability to react to changing information fast or proactively adapt to anticipated change. Technology systems also include computer networks as well as the Internet, WiFi, Global Positioning System (GPS), radio-frequency identification (RFID), and other emerging technologies. A holistic outlook in systems design, encompassing all aspects of the company, allows better categorization, storage, and use of information as well.

Conventional organizations, divided into departments with incentives to locally optimize each department, create their own systems and nomenclature in a proprietary fashion. Over time, each department, group, or location may continue to diverge from each other as they continue to pursue localized strategies in the management of their zones of influence. The company may attempt to create common platforms, now and then, using common systems such as enterprise planning and management systems. Consultants are always lurking to implement systems in big companies as such "transformation projects" utilize a large number of people and run for a long time. By the time the "transformation project" is completed and the company has a shiny computer system in place, such systems may already have become obsolete. Technology obsolescence is fast and furious in this modern era and most users of the "new" system may abandon it the week after implementation and once again pursue localized system strategies to get their work done. This cycle of diverging systems in different parts of the company, the company's attempt at consolidating all systems in a large "transformation project," the immediate obsolescence of the new technology, and the subsequent divergence of the systems used in different parts of the company is common. The inability to use all information in real time (because of segregated technology systems) and the time and resources expended in attempts to tie everything together with rigid technology solutions render companies rigid in the technology system dimension.

Systems: Process

Processes have a hierarchy of activities and a level of repetition. The manufacturing process, for example, may include the definition of the supply chain—procurement of raw materials, storage, transformation, inventory, and shipping of finished goods for a family of products. The company uses this process repeatedly. Since repeating processes provide information advantages that may lead to the minimization of cost or time, maximization of customization premiums, and product enhancements, it is important that the processes are well understood. Many companies spend significant resources to document business processes (using techniques such as process mapping). Such documentation, if it requires much time and effort, may reduce flexibility as the exercise itself may render the processes static. Process definition and understanding need to target the ability not only to operate existing processes efficiently but also change them as new information becomes available. For example, many consulting fads such as "reengineering," "process redesign," "lean manufacturing," and so on, focus on taking status quo processes, documenting it, and making incremental improvements to it so as to reduce cost or time. These are visible projects that allow managers of the firm to take credit in cost reduction and productivity improvement. However, documenting status quo processes and making incremental improvements to it, assuming an unchanging future state, may render these companies rigid. Process flexibility (not process efficiency)—the ability for processes to change so as to adapt to new information in the future—is more important.

Systems: Content

The third component of system is content. This includes systems that define culture, ethics, and morality. These systems need to incorporate the latest thinking as well as the core values of the company. Consistent application of the core values of the company reduces uncertainty in operations. Ability to reflect the latest thinking allows the company to change with the times, thus incorporating the necessary flexibility. The importance of content systems in a company is increasing as the other two components—technology and processes—are more established and less differentiable. In the ethical systems of a company, for example, the principles-driven ethics is much more effective than "rules-based ethics." The continuous stream of fraud that has been discovered in the accounting and financial structure of companies is a constant reminder that "rules" do not create an ethical company. A company that adheres to a principle of transparent communications to all stakeholders is likely more effective than the one that attempts to implement a large number of rules for communications—when, how, what, who, where, and so on. A retrospective analysis of successful and failed companies in this dimension makes it abundantly clear that the content system is a responsibility of everybody in the company and not just the few at the top.

Strategy

The third component of the organization—strategy—includes how the company influences its environment. This includes internal, external, and boundary strategies. Internal strategy entails operations of companies including R&D and manufacturing. External strategy deals with how the company interacts with its surroundings such as competitors and regulators. Boundary strategy describes its interactions with customers, suppliers, partners, and collaborators. Well-implemented strategies allow the company to manage future uncertainty better. For example, competitive strategies include understanding the uncertainty in the behavior of existing and future competitors and designing flexibility into the company's products and services so as to compete effectively whatever the future outcomes are. Companies have to move away from traditional analyses that focus on direct competitors on known dimensions and include the possibility of competing and converging technologies and associated uncertainty. Often, the problem to solve is not how to position the company's products and services against competitive offerings in terms of brand, price, quality, and availability but what emerging technologies and ideas may make it obsolete. So, the true competition for contemporary companies are future companies and yet to be invented technologies.

In finance, the somewhat murky understanding of how shareholder value is created in modern companies has led many astray in the crafting of strategies. For example, one instinctual reaction to sagging stock prices of any company is the acquisition of another company. "Synergies" will be invariably mentioned in the press conference conducted right after the "announcement" by the leaders of the two companies. They may describe how they will remove overlapping functions, locations, people, and machines (virtually everything except perhaps themselves) in an attempt to instantly increase shareholder value. This has been a very fertile area for investment bankers, masters at the design and execution of transactions for the benefit of all. If any shareholder politely points out that the acquisition has reduced the value of the acquirer by X, increased the value of the target by X′, and X′ is less than X, the leaders will point out that investors do not know how capable the managers are in extracting "future value" from the combined company. Empirical data are very strong that mergers and acquisitions (M&A) do not increase shareholder value. However, M&A is a predictable response of most struggling companies. For managers of these companies, M&A provides a useful distraction and for the investment bankers, a nice pocket change. The target's shareholders may also be happy at the cost of the acquirer's shareholders as typically "a premium" is paid over the target's price. Heavy analysis by the investment bankers may have "justified" the premium as their compensation is driven by the amount of the transaction and nothing else. More importantly, in an M&A, both companies lose flexibility. Consolidating physical infrastructure and people may result in the company losing location and skills flexibility. Inherent culture clashes and confusion will bring the combined company to a standstill, unable to respond to the changing environment, further eroding flexibility.

Designing R&D and manufacturing strategies that allow the company to switch quickly among technologies, products, services, and locations to take advantage of new and changing information is more dominant as well. For example, an airline that has consolidated all their operations on a single type of aircraft or a single manufacturer may have gained tactical cost advantages in maintenance but may have lost flexibility in customizing capacity to demand and may have increased the overall risk of the company. Internal strategies pursued by the company have to be consistent with other strategies as well. For example, an automobile company pursuing collaboration with another company internationally has to consider the alignment of manufacturing capacities, types and configurations, currency effects, and interactions of financial and real risk.

In summary, the structure of information, technology systems, and internal strategies are more tangible for the company. Being tangible does not mean they are more or less important for the company. It just means that these components are much more visible for everybody involved. Thus, a larger number of people are involved in the design and maintenance of them. On the other hand, structure of human resources, content systems, and external strategies include intangible elements and are more difficult to understand and influence. Infrastructure, process systems, and boundary strategies are in between. The maturity and scope of the organization can be assessed based on how much importance it places on these aspects. In general, visionary organizations with significant scope will have a lot of focus placed on the intangibles. Conventional organizations focus on the tangible aspects of the company and boundary interactions.

4

Structure

In the last chapter, I defined the three components of an organization—structure, system, and strategy. In this chapter, I dig deeper into one aspect—structure. It includes human resources, information, and infrastructure.

Human Resource Structure

Let's start with the structure of human resources. From the earliest organizations formed by humans to today's modern companies, the human resources structure has remained largely static. Typically, these structures followed a hierarchy where command and control flowed downward in a predesigned and systematic fashion. Early on in human migration, such structures provided migrating clans a way to operate around a single individual or group of individuals who had extra knowledge regarding the terrain, food and water sources, and animal routes. Since such information was gained through experience, the elders in the clan occupied a position of importance. Since knowledge was transmitted only orally, the clan leader then had the option of when and to whom to transfer knowledge. Such transference occurred close to the end of the leader's life, as there was no motivation to do it earlier since such a transfer would have reduced the importance of the current leader. The current leader also selected somebody from his own close circles (if not a direct descendent) so as to keep some level of command on the clan even after the knowledge transfer.

Since the happiness of the clan leader was largely driven by his own well-being, a structure that incorporated the leader at the top of the hierarchy was personally optimal. For his own safety and ease of management of the clan, the next level down in the hierarchy typically contained close associates who took orders from the leader. Through the delegation of certain activities that did not require divulging knowledge, the leader also kept associates happy as they commanded power over subsequent levels in the hierarchy. It was also to the leader's advantage to have close associates specialize in different activities so as to minimize the probability of any individual gaining a holistic understanding that may result in challenging the leader. Such an outcome would have threatened the leader's status as the universal repository of knowledge.

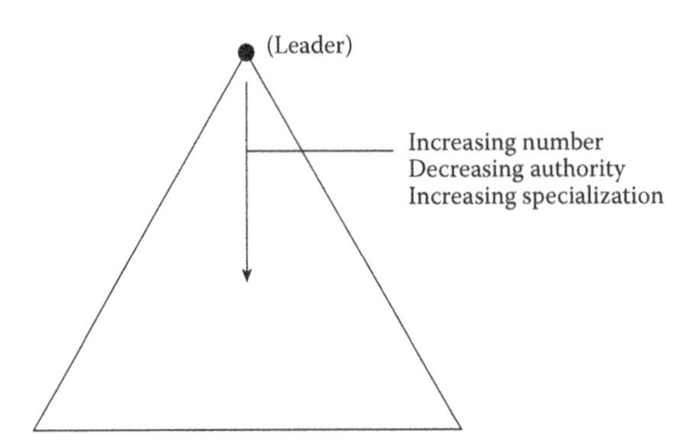

FIGURE 4.1
The pyramidal model of human structure.

This initial structure of human resources has largely remained the same through the ages—a command-and-control structure with fewer people on top and more as you progress downward. At the apex was the leader who integrated all activities and knowledge. I call this the pyramidal model of human structure (see Figure 4.1).

The pyramidal model is efficient in accomplishing prespecified activities. For example, the Egyptians used the pyramidal model, literally, to build the pyramids. The output was clearly defined up front and remained unchanged for decades—a resting space for the leader after death to assure transfer to the after-life and do so with such grandeur so as to surpass the ancestors' attempts at the same. This output had virtually no uncertainty as long as the leader remained alive. When uncertainty is not present in the expected output as well as the environment, a pyramidal model may be quite efficient. In the construction of the pyramids, the jobs were specialized—with more specialization as you moved down the command structure. The activities were repeated for decades—quarry stone, move it to the construction site, and lift it into position.

All through human history since the ancient civilizations, the pyramidal human structure has been the dominant way to organize. This was true for kings as well as religious leaders. The advent of the Industrial Revolution in the eighteenth century resulted in organizations for profit following the same template. For example, automobile manufacturers seeking higher scale and lower transaction costs integrated vertically and managed the entire company with a pyramidal structure. Henry Ford's invention of the conveyor belt was a formalization of low-level specialization in a systematic pyramidal structure. Early on, the owner occupied the leader position and there was no conflict between her happiness and the company's profitability. Later when the owner was forced to appoint agents (managers) as leaders of the organization, the managers' needs conflicted with the maximization of profits and this created a whole slew of problems for the owners.

In designing a modern human structure for organizations, capable of thriving in high levels of uncertainty, I first identify the attributes needed for a successful organization. First, success in the modern economy requires high levels of innovation as it is the most important lever of flexibility. The organization has to grow and nourish the ability to solve problems and take advantage of opportunities as new information arrives randomly. Thus, the design of the modern organization has to optimally encourage and sustain innovation. However, it is impossible to determine beforehand the characteristics of people needed in the organization, to be optimally innovative. This is in marked contrast to the pyramid makers and automotive manufacturers who could clearly define the attributes of workers needed for meeting their production goal, one that was largely static.

If employee characteristics cannot be defined precisely, the status quo human resource processes to recruit and retain employees will not work. Currently, most companies create a "job description" for a new hire. The job description clearly states what the organization expects the employee to do and in some cases not to do. The specification template for the employee also includes a requirements section that details required education and experience. By clearly defining what is needed and who can apply, human resources departments reduce the pool of possible applicants. This is the same process followed by the pyramid makers—who may have required a mason under the age of 25, who has done at least two years of stone shaping before. Similarly, the automotive manufacturer may have sought a mechanic with at least five years' experience in heavy-duty transmissions. Today's service companies follow the same templates. For example, a financial services firm may seek an analyst who has 10 years' experience in the management of risk in fixed income securities, with an MBA from a top tier school. In all these cases, the "job" is defined precisely and then the attributes of the worker who can conduct the job is prescribed. Indirect proxies such as education and experience are used in further defining a set of attributes that are necessary. In many cases, irrelevant attributes such as gender, age, race, and physical dimensions may also be used in the hiring decision, although they are illegal in many countries.

Both high levels of specialization and the pyramidal human structure introduce rigidity and diminish the organization's ability to manage uncertainty. In the pyramidal structure, the success and survival of the organization depends on the actions taken by a small percentage of people at the top of the pyramid. Uncertainty creates unpredictable shocks in the organization and such shocks cannot be effectively managed by a small number of people however smart they are. Integral to the pyramidal structure is a higher level of specialization in lower levels of the organization. Specialization has been synonymous with efficiency from the early days of human clans to pyramid makers and automotive manufacturers and finally to contemporary organizations. Since specialization is a characteristic of the lower rungs of the pyramid, there are only very few in the organization who are routinely integrating the actions of multiple disciplines.

In the clan structure, specialization would have meant that some people acquired special skills such as the killing of animals with an accurate throw of a weapon. Some others may have focused on identifying the presence of animals through smell and other signals and yet others were responsible for the storage and future retrieval of food items. Pyramid makers and industrial companies maintained the same structure requiring people to specialize in various activities. By doing an activity over and over again, each member can get better at what she is doing and thus increase efficiency. This also meant less distraction as everybody knew what to do, uncluttered by the complexity of the end goal. Leaders of the pyramidal structure integrated the various activities to accomplish the end goal, while the participants are either unaware of the goal or were kept in the dark by design. Organizations that depend on a few people at the top of the pyramid for the integration of all specialized activities below them are vulnerable to shocks because they will be unable to react fast enough to them.

Specialization also means compartmentalization of risk. Nonperformance in specific activities can be measured, as they are unaffected by the complexity of the organization. For example, the output of a mason or automotive seat assemblyperson can be measured and compared against benchmarks. The industrial companies took it one level further by establishing timing and quantity thresholds for outputs and conducted performance management based on them. If the production quota is not met, employee wages and management bonuses can be reduced, providing a powerful disincentive. Modern financial institutions conduct risk management in a segmented fashion as well, with various departments conducting different risk management processes. In cases where the overall risk of the organization is not the sum of the risk of the departments, because of the nonlinear interactions among them, the pyramidal leaders were unable to integrate this complex set of risks. In many cases, the pyramidal leader was simply unaware of the total organizational risk and led the organization to bankruptcy. In the recent financial meltdown, well-known firms had departmentalized risk controls that were kept within thresholds individually. However, the leaders of the company were either unaware (or did not care) about the overall enterprise risk due to the nonlinear interactions between the departments. Although the house of cards may have looked good from outside, once a card fell, it created a series of chain reactions that took the entire company down.

Leaders of pyramidal organizations may also inherently believe that their presence is of fundamental importance to the company. In a famous interview of a senior leader of a failing automotive company, a television reporter asked him if he would consider any management changes given the dire straits the company is in. The executive responded with confidence that the company needs the best and the brightest in its time of peril. He went on to add that the current management team was the best as they have been around for a long time. Unfortunately this was the same period in which the company's stock price lost 95% of its value. The market for pyramidal leaders is also a small one and it is an exclusive club. This idea is perpetuated by the leaders and the recruiters who manage them. Recently, a financial services firm that was at the brink of bankruptcy sought to replace its CEO. There were only half a dozen people in the entire world qualified to do the job. A reporter on business TV remarked, "The company tried its best to recruit somebody, but it is so complex that only a handful of men can even think of doing that job." The ones in the "approved list" recently ran companies of similar size to failure as well. After recruiting a qualified CEO, the company required many bailouts from the government and is still at the brink of bankruptcy. So, once somebody climbs to the apex of a tall pyramid, she can jump from pyramid to pyramid regardless of her competence or her ability to lead companies successfully.

The other commonly seen feature of the pyramidal structure is defined layers with information opaqueness between them. As shown in Figure 4.2, each layer is set up such that the participants in that layer interact with each other more than those in other layers. In certain companies, this literally happens in large buildings with different floors. Each floor may have a defined layer. Typically, the top layers reside in higher floors closely matching the classical pyramid. Each layer and floor may be decorated differently with the top layers commanding a higher level of opulence. In a recent case, one of the leaders of a failed company spent over one million dollars decorating his top floor penthouse office as the company was going bankrupt and was being bailed out by the government. Such arrogance is commonplace in pyramidal organizations where the leader does not have much accountability for these types of actions. It is ironic that while the board and shareholders were biting their nails at the imminent collapse of the firm, the leader of the company was out shopping for a better carpet. Some believe that the scrutiny of such

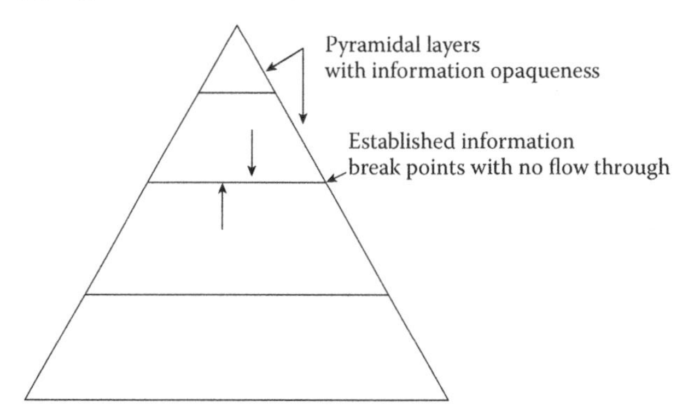

FIGURE 4.2
Information opaqueness between levels in pyramidal human structure.

"minor expenses" are not warranted but as the recent financial crisis taught us, they are symptomatic of a larger problem in organizations. It is not the amount of money spent decorating personal offices that is important but rather the decision to do so. It clearly shows the character of the person at the helm and it may signal attributes of the culture that permeate the company at all levels.

Occupants of a layer (floor) may have meetings and share information but seldom do so with other layers in the organization. Meanwhile, each layer in the organization defines its own operating modality and information exchange. Historically, both the clan leader and the pyramid makers had reasons to curtail information flow. For the clan leader, information flow down the organization may have created challenges to his dominance. If everybody knew better hunting routes and water holes, the clan leader's grip on the organization may weaken. For the pyramid makers and industrial companies, the situation is similar as information opaqueness helps sustain the power structure within the organization. With few choices, the owners (shareholders) of the organization typically give in to the arguments made by the pyramidal leader about why such information opaqueness is good for the organization. Since the shareholders run the risk of loss of critical information if the leader were to leave, the leader holds power over them and not the other way around. So, owners of the firm may suboptimize the selection and retention of the leaders of the firm as they prioritize the minimization of tactical operational disruption over long-term viability.

Information opaqueness increases when the communication between layers decreases. There are many mechanisms that are designed to maintain this information opaqueness such as segregated offices, locations, off-site meetings, and so on. Information generated in the layers tends to remain there due to this phenomenon either naturally or due to the conscious efforts of the participants of the layer to hoard information. Even in cases where the participants in top layers wish to reduce such opaqueness either because they would like to take a peek at the hoarded information in the lower layers or because they would like to increase organizational flexibility, there are structural impediments in most pyramidal organizations that make this difficult to do. Human resources managers have played a gatekeeper role in such companies disallowing any random communications between layers. In extreme cases, e-mail, personnel, and physical security have also played a role in virtually eliminating any unplanned communications between layers. E-mails sent between layers are sometimes screened at the boundaries by a few who are given explicit instructions not to allow random communications. In some cases, security personnel are deployed to disallow employees to meet leaders of the company. In other cases, the executive suites are kept under lock and key, given controlled access, and monitored to virtually eliminate any random communications. All of these actions emerge from the basic idea that the participants in the top layers have all the information to run the company. Historically, it is this same belief that drove the clan

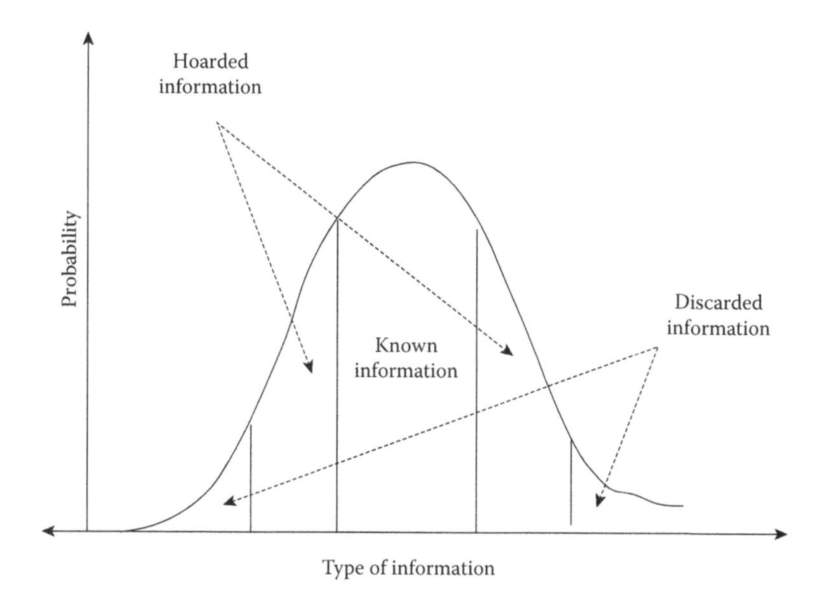

FIGURE 4.3
Information histogram.

leaders in early human societies and later pyramid makers and managers of the industrial companies. Any unknown information was considered a distraction and was avoided at any cost. One way to do this was to eliminate unplanned interactions between layers.

Information hoarding is an underlying behavioral aspect of the pyramid structure. This is a two way process. As information opaqueness is designed into the pyramidal structure top-down, it also provides incentives for information hoarding in the lower layers. Bottom layers keep some information hidden and only release it selectively. There is a subtle difference in the bottom-up and top-down information hoarding behavior. Since the pyramidal leader and the top layers can command information from the bottom layers, they will do so for information that is known to exist. If the bottom layers are in possession of other information, it is dominant for them to hoard it and only release it selectively.

Let's look at information hoarding more closely. Information can be represented in a histogram or probability distribution as shown in Figure 4.3. Most of the known information exists in the middle and is available to many. This known information is the primary currency of transaction between the pyramidal layers. Since both the top and bottom layers of the pyramid know the existence of such information, whenever the top layer needs it, it can just ask for it from the bottom layers. The bottom layers are obligated to provide it and such information can be easily verified. To the left and right of the known information in the histogram is information that is available to the participants of the bottom layers. The top layers are unaware of such information and thus cannot ask for it to be revealed. The bottom layers, thus, are not obligated to provide it and hoarding such information brings certain proprietary and monetary advantages to the bottom layers due to the timing option for revelation.

One advantage of hoarding information unknown to the top layers and known only to the bottom layers is the ability for selective revelation for favors and promotions. In traditional companies, "stars" are identified in the bottom layers when previously unknown information is revealed at an opportune time. If such information is revealed optimally and identified by the top layers, the bottom layer participants may be rewarded with higher compensation or an elevation (promotion) to the layers above. Since it is a right and not an obligation, the ability to reveal information optimally is an option held by bottom layer participants. As the readers may know, it is not optimal to exercise options prematurely. Since the value of the option held by bottom layer participants are driven by the uncertainty in the benefits that can be obtained, it is optimal to hold such options as long as possible. However, these options are proprietary to the information hoarder and cannot be sold (at least not explicitly); there is an optimal time to exercise them. It is also the case that the relevance of the information held by the participant may decline over time and she also runs the risk of accidently revealing it or others capturing the same information. So, such information hoarders will delay the revelation somewhat, especially when there is high uncertainty in the outcome of the information revelation (such as anticipated promotion and increase in compensation) but exercise the option when the anticipated benefits are high enough. Since the benefits accruing to the information hoarder is proportional to the novelty of the information and the possible use for such information to higher layers (and not necessarily to the added benefit to the company), it is common to observe spectacular cases of information arrival in predetermined forums and time points when the information hoarder has maximum visibility. In any case, such information hoarding and delayed revelation reduces organizational flexibility and reinforces layered pyramids and information opaqueness.

Equally damaging is the discarding of information in the tail of the information histogram. To the left and right of the hoarded information is unusual information that has very low probability of occurring. The top layers are unaware of such information. More importantly, the designed transactions in the known information and the active hoarding of adjacent information lead to the bottom layers discarding such low probability information elements. Since this type of information was never used in the company—neither in direct communication with the top layers nor in selective revelation—the bottom layer participants do not realize the value in it or it may be risky for them to collect and communicate it. For example, the oddity in the data of a clinical trial by a pharmaceutical company may have implied a higher probability of failure of the drug in the future. The data are very obscure and only few scientists in the bottom layers of the company may have seen it. They may simply discard it as an outlier. The other possibility is that the collection and communication of this type of information is risky for the participant. Since this type of information may not be used with optimal timing (it has to be revealed immediately), it may be risky for the lower participants to collect and analyze it. If the information is hoarded and not revealed, future revelation will show that the participant had the information beforehand and decided not to communicate it. Since R&D is an expensive activity, this may spell trouble for the owner of the "hot" information. This type of low probability information is also more likely to lead to innovation. The loss of this information due to the pyramidal structure and related information opaqueness, hoarding, and incentives thus increases organizational rigidity and reduces innovation. The value lost in discarded information is very high for companies in innovation-driven industries such as life sciences, high technology, and financial services.

Now consider an alternative human structure I call the SOUL (self-organizing uni-layer). In the SOUL structure, there is no singular leader and no defined hierarchy. The SOUL is a network of people who form the organization by self-subscription. That is to say, in the SOUL structure, the company does not hire employees, but rather the members of the SOUL voluntarily join the "SOUL." The SOUL may have a macro-objective such as "invent and market technologies that reduce the consumption of power." The members of the SOUL are naturally interested in the objective of the SOUL. The members thus may not have any prespecified education or experience. The SOUL may be financed by a group of shareholders (owners) who also are fundamentally attracted to the objective of the SOUL. The owners do not have named agents (officers who reside on top of the pyramid in contemporary structures) but all SOUL participants will be considered agents of the owners. In situations where the SOUL has not sought external capital, all SOUL participants may be owners as well.

The SOUL structure thus does not have a defined shape, nor does it have a defined command-and-control structure. It is a dynamic structure, willing and able to change shape and interactions as the situation and the environment demand (see Figure 4.4).

The SOUL structure has the following characteristics:

1. SOUL participants self-subscribe. This eliminates all common human resources processes such as hiring and retention, job specifications, and selection dimensions such as age, education, experience, and other tangible attributes. The SOUL thus assembles a group of people tied only by its objective. They may bring diversity in all dimensions. If the SOUL starts with low diversity, the SOUL members can increase diversity quickly if they believe it is beneficial. Unpredictable future shocks to the organization will be handled by the SOUL as a whole. The SOUL is not vulnerable to the whims and incompetence of a few at the top of the pyramidal structure of typical companies.

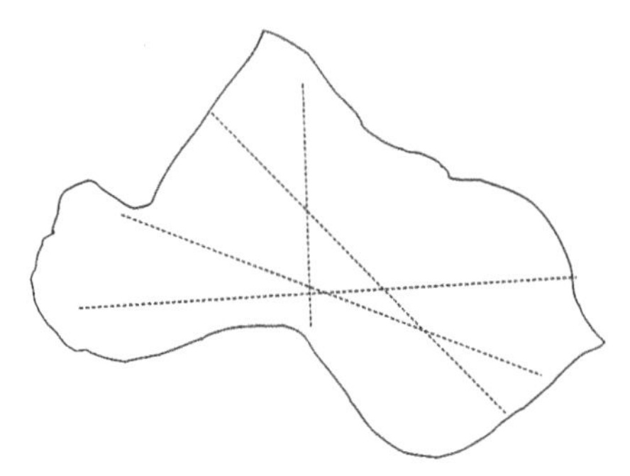

FIGURE 4.4
Self-organizing uni-layer (SOUL) structure with undefined shape and unplanned interactions.

2. The SOUL manages itself. Since the command structure is absent, decisions can be democratized within the SOUL. For example, all SOUL members can vote on any decision the SOUL makes. Technology allows this to happen without high investment in time and effort. Since SOUL-type companies will be small, implementation of democratized decision making is easy to implement. The input into decision making will be voluntary for SOUL participants, and they will engage using social networking technologies if and only if their inputs are important for the decision under consideration.
3. The SOUL compensates itself. Since SOUL participants are either owners or direct agents of the owners, the compensation of the SOUL participants will be directly related to the success of the SOUL.
4. By removing both traditional managers and compensation structures from the SOUL, there will not be any incentives to increase the size of the SOUL to gain titles and promotions so as to climb the pyramid (as compensation is typically driven by titles in companies) or to simply stroke one's own ego.

The SOUL thus removes the inefficiencies that exist in current companies related to hiring, retention, firing, compensation, and incentives. At the same time, it adds flexibility to the organization to cope with uncertainty.

For example, consider a SOUL that is in biotechnology. Most biotechnology companies are formed around a defined intellectual property (IP) position. At the formation of the company much effort is spent in appointing the CEO, CFO, chief scientific officer (CSO), and so on; each of these titles have some preset definition and expectations. This act introduces rigidity into the company and impedes its ability to advance the IP and enhance value. All the designated C positions specialize in something, and except for the CEO, who is expected to integrate the activities of all participants, nobody in the company is now directly accountable for the increase in the value of the company.

Each is given a set of actions (such as managing finances and conducting clinical trials), but none other than the CEO has the direct responsibility to meet the objectives of the company or increase overall value. Job descriptions are clearly defined and many of these may have nothing to do with what the company is trying to accomplish. If the founder (who initiated the IP position) elected to create a SOUL instead, this would mean that anybody who is interested in joining the SOUL will be interested in its objective—say, "find a cure for Alzheimer's disease." If the SOUL seeks capital, the capital provider will also be fundamentally interested in the advancement of the objective of the company and less animated about the CEO and the CFO. The SOUL biotechnology company will be a group of people with different backgrounds and skills tied together only by an interest in the end outcome. They will have substantial stake in the company's success and will be largely compensated by the end outcome. This is against the conventional notions held by venture capitalists, who seek to establish the "management team," first and then think about the objectives. The idea is that if the management team "has done it before," they are well equipped to do it again. As can be seen in many situations, the end game in the biotechnology business is largely driven by luck and thus past success is never a good indication of future success as far as the management team is concerned. It is the ability of the participants in the company to work together and solve problems that increases its ability to succeed. The myth of spectacular managers creating successful companies single-handedly in the presence of incompetent employees is perpetuated by those who benefit from it. Over time capital providers, recruiters, and ultimately shareholders begin to believe in it even though there is no empirical data that supports it. Those who analyze the problem can clearly see that the biotechnology industry is dominated by technical risk. In spite of the best efforts of those who are involved, the company can fail. It is equally possible that the company succeeds in spite of the employees and the incompetence of the management.

The SOUL is also able to dismantle itself without significant cost. Biotechnology companies go through many rounds of private funding—each capital provider seeking certain returns within a window (and less interested in the ultimate objectives of the company) eventually culminating in external financing, aided by investment bankers. With high-priced leaders on top of the conventional pyramid, the "track of capitalization" of the company is fixed. Such financing is done regardless of the advancement of the company's objectives or information that becomes available over time. External financing is not done based on the value of the company but rather the templates are followed over and over again by the capital providers, who are more interested in the transaction and less interested in what the company is all about. In some sense, the leaders on the top of the pyramid and the capital providers "run" on a set course that is disconnected with the company's objectives and prospects. In many cases, the most optimal action for the SOUL will be to dismantle itself when it realizes that the idea it is following is not feasible. The SOUL participants can disengage and move to another SOUL that is of interest to them. The exercise of this "abandonment option" is more optimal when the SOUL has the flexibility both to dismantle itself and the SOUL participants have more flexibility to reorganize into other SOULs. In today's companies, early capital providers and the management have an incentive to advance the company to an "exit" even if it is clear that the end objective of the SOUL is not achievable. In many cases this is wishful thinking as the concept of "sunk costs" is difficult to accept for many. This leads to incremental investments to prolong the company that does not fundamentally change its chances to reach the next round of financing. It also leads to contentious games played between classes of capital providers, vying for larger chunks of the company. While the investors and the leaders of the company put lipstick on the pig and battle against each other for bigger chunks, the pig may be dying.

The SOUL structure substantially removes information opaqueness between layers since no layer exists. Incentives for information hoarding are also removed, allowing the organization to be fully information efficient all the time. Since information is considered to be something that is shared across the organization, the chance of losing the "tail" (unusual) information also diminishes. The SOUL participants thus can focus fundamentally on the tail information as the identification, sharing, and using of known information is virtually automatic. This unusual information is the primary component of innovation and increased focus on it will help the SOUL enhance its ability to innovate, further adding to flexibility.

A community of interconnected SOULs also increases system flexibility. At the limit, one can imagine an ecosystem of SOULs that is constantly organizing itself based on the latest available information. In such a system, nobody is trying to keep jobs and titles—an activity that introduces rigidity in current companies. Each SOUL participant will have an incentive to engage with the SOULs that are most suited for her skills. Participants will also have an incentive to disengage when new information indicates that either the SOUL's objective cannot be met or other SOULs are a better match for their skills. This sounds like a bad idea to those used to systems and jobs that demand loyalty in the pyramidal structure. Loyalty simply means that the leader in the pyramidal structure expects subordinates to stay within the pyramid till they are ready to "depart." That is not necessarily loyalty but an idea invented by the leaders of the pyramid to avoid chaos in the ranks. In cases where such loyalty cannot be demanded, pyramidal leaders and owners may bestow golden handcuffs and stock options on employees in an effort to tie them down. One mechanism invented by pyramidal leaders to "keep employees in their jobs" is "employee stock options." This was created under the guise of aligning incentives of the employees but works to increase switching costs for the employees and thus reduces the frequency of switching to other jobs. The boards of conventional companies have also invented many "golden handcuffs" for top executives in an attempt to keep them from switching. So, even if the leaders do not like to be at the company, such golden handcuffs will keep them there, possibly destroying productivity and decreasing shareholder value. Most of the instruments deployed to discourage switching in contemporary companies work against the company by keeping employees not well suited to the company's objectives inside the company forever. Compensation schemes that focus on eliminating or diminishing the opportunities for employees to move to more productive companies reduce the overall flexibility for the organization and the economy.

A dynamic engagement and disengagement of participants with various SOULs in a community of SOULs also may sound like chaos to traditionalists who may fear that "people may jump ship" at will. Systems that are fully dynamic—in which only SOULs exist—can be shown to be a much more efficient system in this regard compared to a fully static system of existing companies. In a static system, the employee and the employer incur high costs when the employee switches from one company to another. For the employee, this takes search costs, location change, retraining, and other transition costs. When such costs are high, most opt not to switch from one company to another and continue in the current suboptimal "job." The current job may not be the best for the employee either because her skills are not in sync with the job or the job may not be adding value to the company. For the employer, the loss of the employee creates costs in the search for a replacement, training, and transitional loss of productivity. The employer thus attempts to keep the cost of switching high for the employee. This introduces inflexibility for both the employee and the employer and they both settle for a defined and less uncertain but suboptimal outcome.

The SOUL structure, however, does not guarantee the assemblage of the best group of people to achieve the company's objectives. The SOUL thus has to keep the option to reject an application for subscription. For example, if an individual starts a SOUL based on a new idea or IP position, she will have the option to reject any subscription to the SOUL upfront. Once the SOUL starts to grow, each additional member of the SOUL will be accepted democratically as all SOUL members will have an equal say in the acceptance of new members. Since the SOUL members know what skills and characteristics best complement the current members and help the SOUL achieve its objectives, the combined decision to accept or reject a subscription will be optimal. In today's companies, the top executives make hiring decisions for their own departments. Thus, the CFO will hire finance professionals and the CSO will focus on hiring technical talent. In such a situation, individuals tend to replicate themselves in the departments by seeking similar attributes in hires as their own, such as education, orientation, and skills. In older companies, such attributes also include color, gender, and age. This results in a set of clones in each department with individuals similar to each other. Such an organization will introduce rigidity in human structure and will be unable to succeed when the environment changes.

If the SOUL starts with a single person, decisions are by definition democratic. Once it reaches two people, the SOUL may require consensus in further hiring. Some may argue that this will lead to deadlock and the SOUL will lose its ability to grow. Since the members of the SOUL are the primary contributors of success of the SOUL, if democratic processes lead to deadlock, it implies that the SOUL is not finding the right subscribers. Growth, although important, is not a necessary condition for success for companies and so it may be better to wait. Many start-up companies fail not because of ideas or ability but because it hired the wrong people, perhaps following templates provided by the capital provider or perhaps based on common beliefs held in the industry.

Since the success of the SOUL largely depends on its ability to assemble a group of people able and willing to progress the company's objectives, it has to actively seek to increase the pool of subscribers. But unlike firms who seek employees at recruiting events and through headhunters, the SOUL should attempt to provide as much information as possible about the company's objectives and culture to the general public. Rather than selecting recruiting locations such as college campuses and employing headhunters, who literally hunt for employees based on set specifications, the SOUL should provide information using widely available mechanisms such as the Internet and the media. Since most SOULs also have location flexibility, such information should target the entire world rather than specific locations and countries. By increasing the potential subscription pool, a SOUL has a higher probability of assembling the best group of people willing and able to work toward its end objective.

The SOUL manages and compensates itself. Since all SOUL participants have an unambiguous goal of achieving the SOUL's objective, it can eliminate all management-related loss of productivity. The appointment of agents (managers) by the owners (shareholders) from the pyramidal makers to auto manufacturers have created many distortions in companies. For example, in the pyramidal structure, power is measured by the number of subordinates. Since compensation is also correlated with power, the manager's primary goal will be to increase the number of subordinates. This makes the manager either decide to grow the company quickly organically or acquire other companies. The bigger the company, the higher the manager's power and likely the higher the compensation. Shareholders have also used the size of the company as the proxy for the manager's ability. Those who manage larger companies are considered a special breed and provided compensation at a much higher level. In the larger ecosystem of current companies, those who have experience in large companies have been given a premium. Thus, the manager's incentive to grow the company is not only for her current position but also for her career. As can be seen from the constant recycling of large company CEOs, entry into such a small club is immensely valuable for the pyramidal leader.

The SOUL is similar to a single cell organism. There are zones of specializations but none of these are given more importance than the other. For example, a single cell has a nucleus, membrane, cytoplasm, ribosomes, mitochondria, lysosomes, vacuoles, and so on. All these specialized attributes need to come together for the single cell organism to function. In the absence of any one of them, the cell will cease to function and hence it is impossible to give importance to one over the other. Some single cell organisms, such as a bacterium, have been the most successful on earth and in many ways still dominate the world. Similarly, all specialties are equally important in a SOUL and it cannot operate in the absence of one. Thus, the compensation structure of the SOUL will be flat. The difference in compensation between the participants in a company is an indicator of how pyramidal (or less SOUL like) the company might be. Similarly, companies where certain functions, such as marketing or finance, dominate also are unlikely to reflect SOUL-like features. I return to this discussion in Chapter 8 as I put together a diagnostic kit for existing companies.

In raising capital, SOUL-type organizations should consider making use of the Internet to broadly reach potential investors. As demonstrated in the recent U.S. presidential elections, a larger group of people making smaller contributions can be a powerful motivator for success. Similarly, rather than seeking capital from traditional sources, SOULs may raise capital on the Internet using an auction. In this case, the SOUL has a much higher probability of assembling the right type of investors who are more interested in its objectives rather than financiers looking for exits and hard metrics for expected returns. A large Internet company considered this possibility during its initial public offering (IPO) but did not follow through and succumbed to the pressures of investment bankers and industry standard templates. There are no technology impediments today that prevent companies from raising capital on the Internet. The only barriers that exist are those raised by monopolistic institutions and regulations that assist them. It is high time that companies broke the shackles of inefficient intermediation as practiced today that result in a high percentage of the value of the company siphoned away by intermediaries.

It is possible for a company to acquire SOUL-like characteristics over time. It will have to both shed participants who do not fit the SOUL structure and fundamentally transform its compensation and incentive structure. It is a long transformation process and depends very much on status quo size, culture, and employment contracts. In many cases, it is possible that the existing company's human resource structure has lost all flexibility and improving it incrementally is no longer possible. In this case, the company should consider splitting itself into smaller ones or starting over from scratch.

To increase its value, the company has to both increase its ability to innovate as well as introduce flexibility to alter uncertain future cash flows favorably. The ability to innovate largely depends on the human structure. It depends both on the organizational structure as well as the incentives that exist to innovate. In pyramidal structures, innovation was a distraction. None of the participants had any incentive to innovate. If an innovation resulted in a reduction in time to conduct a task, it just meant that either a larger pyramid or a new pyramid will be commissioned. From the perspective of the participants, this did not add any additional utility. The conveyor belt model of manufacturing in the automotive industry was similar. No incentive was present for innovation. In the SOUL structure, however, innovation will be everybody's business. Any activity that is repeated will be automated and all SOUL participants will focus on fundamentally innovating to meet the objectives of the SOUL.

Information Structure

Now, let's study the information structure in more detail. Information structure is a much more holistic concept than the data structures used in information technology. The definition of the information structure starts from how the information will be used within the company. This is in contrast to most information acquisition projects initiated by companies such as the implementation of enterprise resource planning (ERP) systems. These systems, originating in the 1990s, allowed companies to collect large amounts of data without a good definition of information structure including its use. The idea was to "first collect the data and ask questions later." Both the ERP software vendors and implementers had no incentive to define an information structure for the company prior to implementing the system. Fear of the unknown was used as a method to get fast adoption of such systems. Companies did not understand how they were going to use the data but many were fearful of their competitors gaining an advantage from such systems if they did not implement the same. A wave of ERP implementations happened but the overall return on investment in such data collection vehicles has been low. Any repeating tasks such as accounting, inventory, and payroll can be fully automated using commodity technologies and should not be a concern for companies as they define the information structure.

Information structure of the company starts from how the company uses information to make decisions. The primary information needed for decisions within the company forms the fundamental basis of the information structure. Since decisions by definition are about actions in the present and in the future, the information structure should be able to combine existing (historical) data and forecasts in a common architecture. Since most decisions have to be made under uncertainty, the information structure has to also include uncertainty as a fundamental component. This is in contrast to the current data structures that are deterministic—static and constant numbers (as seen in spreadsheets)—whereas uncertain data can be represented in ranges or probability distributions. For example, if we forecast the price of aluminum next year, we may want to capture the range of possible prices based on demand and capacity uncertainty. In a typical data structure, we may capture the forecast to be $1.0/kg; in a better defined information structure, this may be captured as "between $0.8/kg to $1.2/kg" or more likely in a probability distribution with prices on the x axis and the probability of that price to be the average price next year on the y axis. In such a diagram, we may see that the probability of the average price of aluminum to be $1.0/kg next year is 20%. Once this is established, we can also create confidence intervals. For example, we can derive that the 80% confidence interval for average prices next year is between $0.9/kg and $1.1/kg. Also, a cumulative probability can be inferred. For example, we could note that we are 95% confident that prices will be more than $0.82/kg next year. Here, we are not discussing the analysis of the data but how the information itself can represent uncertainty inherently. Such a forecast may be made using historical data and the analysis of future expectations of demand and supply of aluminum. Demand may depend on many factors such as growth rates in various parts of the world, availability and popularity of aluminum substitutes, technology changes, new products, tariffs, and other factors. Similarly, supply may depend on known reserves of aluminum, production and transportation technology changes, regulations, and other factors. Both demand and supply are also driven by the price of aluminum and the demand and prices of the end products made out of aluminum. All of these factors are uncertain and have to be represented in the information structure as uncertain.

The information structure of a company includes both business and technical information. For example, in a pharmaceutical company, data from clinical experiments are collected to prove or disprove a hypothesis. The technical go/no-go decisions in pharmaceutical R&D are typically made on a therapeutic index such as the therapeutic ratio (TR), which is the ratio of the dose of the drug that creates a toxic effect for 50% of the study population and the minimum effective dose that is needed to create an efficacious effect in the same population. The higher the toxic dose is, the higher the TR, and the lower the efficacious dose is, the higher the TR. The difference between a toxic and an efficacious dose can be very narrow (the therapeutic range or therapeutic window) and so the assessment of these doses is critical for making a decision to go forward with the drug after a stage is completed. In each stage, the drug may be tested in a certain number of subjects or patients. For example, in phase I, healthy volunteers (subjects) take part in the experiment for the assessment of the toxicity of the drug. Similarly, in phase IIA, a group of patients goes through the experiment for the assessment of the efficacy of the drug. Because all human beings are not the same, the effect seen in one will differ from the effects seen in others and thus the information collected is not precise but probabilistic. Generally, the more information available (larger number of subjects and patients), the higher the confidence in the assessment of the TR is and the higher the chance of making the "correct" technical go/no-go decision. Clinical study design has to take into account the null hypothesis (which may be accepted or rejected) and the sample size needed to reach a statistical conclusion. Readers may be familiar with type I (false-positive) and type II (false-negative) errors. Since we start most experiments with a hypothesis and collect sufficient information to either accept or reject the null hypothesis, both of these types of errors are costly mistakes for the company conducting R&D. If a type II error occurs (incorrectly concluding that the drug is not beneficial when it actually is), the R&D program may be abandoned. This is the error of accepting the null hypothesis when it should have been rejected. In this case, a potentially good drug will be abandoned, and the investments made up to that stage become a waste; more importantly, the company will "throw away" a valuable prototype. If a type I error occurs (incorrectly concluding that the drug is beneficial when it is not), the R&D program will progress to subsequent stages, forcing the company to perform larger and larger experiments. This is the error of rejecting the null hypothesis when it should have been accepted. As the "bad drug" progresses through the R&D process, "burning more money," more information will become available. This may result in the abandonment of the drug at a later stage. In some cases, the company may continue to make type I errors in all stages of R&D, taking the drug to market. However, once the drug is on the market, the number of patients taking it dramatically increases, and the information exponentially expands, making the error very transparent. Adverse effects may show up in some percentage of the population, and the actual therapeutic range reveals itself. Depending on the level of toxicity, a number of different outcomes—such as the FDA requirement of sterner warning, a black box label, hospital-based administration, or withdrawal from the market—are possible. Both type I and type II errors are extremely costly for the company. However, there is something much more costly for the company underlying the statistical analysis and that is the assumption of normality in the data. The statistical tests explained above are based on the assumptions that the data collected from clinical tests are normally distributed. This assumption allows the decision maker to discard data that are in the tails of the distribution as outliers. It is quite possible that it is the outliers that make or break the company. So, the current practice of the capture, analysis, and hypothesis testing of technical data are designed to help companies make mediocre and perhaps safer products and discard groundbreaking products. By incorporating technical data with rich uncertainty without underlying normality assumptions, companies can create products that are truly innovative. These groundbreaking products and ideas are the ones that increase the value of the company. Churning more mediocre products at a faster rate is a prescription for disaster for today's companies.

Databases underlying current information technologies are designed to collect and report static and constant data. In most cases, decision makers may be aware of the uncertainty surrounding the data but they are often powerless to do anything else as the available data are always in a deterministic form. To get around this problem, analysts will take such data and run scenarios on ad-hoc tools such as Excel® creating many numbers—all of them having a singular value in each scenario. Since data has to be "processed" in this fashion by many different departments in the company to aid decision making, many different analytical tools, techniques, and formats may exist. Once data are taken out of information systems, further processing of the data moves along disconnected paths and the end outcomes are not stored or shared as they may not fit with the "data architecture." Explosion of computing power has increased the ability of information technologists and analysts to collect and process disconnected and constant data, breeding confusion in the organization.

In today's companies, many departments that are responsible for the analysis of the available data to make a variety of decisions such as raw material procurement, production, pricing, and logistics may exist. Although the tactical efficiency in data storage and processing has increased in companies, the same phenomenon may have led many companies toward a path of lower decision quality. Decision quality means the ability to use all information available (including the uncertainty surrounding them) in making a decision. If the inputs have uncertainty and cannot be represented in a deterministic form, any decisions reached by using such data are of lower quality. Further, the tremendous increase in data storage may have reduced the decision maker's ability to use relevant data, further eroding decision quality.

For example, a decision maker may want to know whether it is profitable to enter a new market segment with an existing product. Internal databases may provide data on cost of manufacturing of the product, external databases may provide the price of competitive products, and customer surveys may provide market share estimates. Based on the current information, forecasts may be made that provide precise estimates as to gross and net margins, revenue growth, and overall profitability. An internal rate of return calculation may be done and that may be compared to the average cost of capital of the company. All these are shown to be precise and if the rate of return is lower than what the decision maker expected, she may forgo the opportunity. There is a famous story that a well-known strategy consulting firm advised a large telecommunications company not to enter the wireless market in the 1980s because the target market was small. The strategist calculated penetration rates, use rates, pricing power, cost of provision, and other aspects and found that the profits the company can reasonably derive from entering the new market are low. Unfortunately, this firm lived to see the day its competitive advantage slipped away as nimble competitors invested and garnered large profits from an uncertain and evolving wireless market. This is but one example of the perils of utilizing unchanging data in making future decisions. For individuals, the increase of data on the Internet may be eroding decision quality as companies compete to gain priority in search engine listings. Mindless crawler programs, searching large amounts of noisy and low quality data, attempt to categorize them. In most cases, the priorities are driven by marketing incentives rather than the relevance of information sought. The provided information also does not show any uncertainty that may be present.

Why does one need to capture uncertainty in the information structure? This goes back to earlier discussions. Managing uncertainty is fundamentally what businesses do. If there were no uncertainty, we could return to the pyramid makers' philosophy of production and this time we can automate everything. If there is no uncertainty, machines will be able to do all the work and there will be no need for humans in companies with set goals. Decisions made on unchanging data can also be automated and the premium demanded by decision makers in management will not be justified. Consulting firms have virtually automated many of the analyses they conduct and the template library they have created contains automated analytics that allow them to "plug and play." The replication and recycling of automated analytics from different industries and companies may have contributed to the loss of competitive position for many companies in the United States. Companies have to first design the information structure that captures the uncertainty in the data elements so that it can be effectively used in decision making.

Also needed are new technologies for the capture, storage, and retrieval of uncertain data. Information structure should capture just enough information that spans the entire decision spectrum of the company. This is in contrast to the current practices of collecting any and every piece of data in the hope of possible use in the future. The approach of collecting the data and then finding some use for it wastes valuable time and resources in the company and renders it inflexible. In contrast, the information structure should be designed from first identifying the decisions and then defining the minimum amount of data that will help make the decisions. Information structure also has to be dynamic so as to change with the company and the decisions it makes today and may make in the future. Static databases are unlikely to satisfy the need for a dynamic information structure. This may require an open-source view of information structure. Since information is valuable, incentives may exist for interacting organizations to modify, append, and delete aspects of each other's information structure. In the conventional world of "proprietary data," this may appear unacceptable for fear of loss of intellectual property. However, the definition of intellectual property is also changing as knowledge is now driven by skills rather than proprietary access to information. This has implications for large companies currently maintaining competitive advantage because of a large number of patents, aided by the scale in the manufacturing of patents through an obscure process conducted by the current patent offices. "Patenting" today is more about legalese than true knowledge and this will change in the future. Proprietary advantages are also driven by scaled access to data for large companies. Since both the need for data as well as the characteristics of the data needed are changing, companies built on the scale of the capture, processing, and reporting of data will probably cease to exist in the future.

Information structure thus needs to be minimal (just enough information to reach decisions), be designed top-down (decisions to data, rather than the other way around), be dynamic (able to change and be changed as new information arrives in the company and in the world), and contain uncertainty as an integral part of the information. "Fuzzy" databases containing not only the data but the uncertain and dynamic characteristics of the data need to be designed. Information bases may exist in a "cloud" (outside the boundaries of a company), may be shared within a network of interested parties, and may be dynamically changed by all parties. Such an information structure will provide maximum flexibility for the company and its network partners to deal effectively with uncertainty. Shared information clouds are not industry specific. They can aid exception-based management, able to alert decision makers to specific problem and opportunity areas as they surface. For example, initiation and spread of diseases can be identified early if data are shared across consumer products, pharmacies, and hospitals. Similarly, the probability of a terrorist attack can be defined better from stock market price patterns, travel changes, and hoarding of consumer goods. Although some government agencies are currently using such techniques, they are not currently available for commercial entities and individuals. This will change in the future and companies designed to take advantage of holistic and uncertain information will have an advantage.

To fully define the uncertainty in all aspects of information and the possible interactions among them, let's study the type of critical information used in decisions in a pharmaceutical company. The pharmaceutical business currently is largely based on creating chemical agents to effect a specific action in a complex biological entity. These new chemical entities (NCEs) can be designed to have specific actions on cell surface receptors that modulate the cellular signaling pathways controlling cellular function. Pharmacology is the science of how drugs affect biological entities. It has two aspects: pharmacokinetics (how the body affects the chemical) and pharmacodynamics (how the chemical affects the body). Since the NCE is "new" by definition, its pharmacokinetic properties (absorption, distribution, metabolism, and excretion) are not precisely known. The manufacturing process of the NCE also has to be invented from available chemicals (some of these processes can have as many as 15 manufacturing steps of reactions starting from what is called alpha raw materials (common chemicals). Each manufacturing step represents the manufacturing of an intermediate chemical from existing ones, which then feeds into the subsequent step. Also, one can only guess the pharmacodynamic properties of the NCE and its hypothesized effect on cell receptors and how such effects ultimately translate into efficacy (beneficial effect) and toxicity (bad effect).

The chemical has to be tested in complex biological systems, but the effects of the chemical cannot be precisely predicted. This means that all through the R&D process we are dealing with various types of uncertainties, and the decisions regarding the nature, timing, and size of manufacturing and testing activities have to be made in an uncertain environment. These uncertainties can be largely divided into the following four categories.

First, let's consider the uncertainties around costs. The cost of conducting an R&D program depends on the people and materials needed to manufacture and test the prototype. The prototype here typically is an NCE or an NDE (new device entity). The "people costs" (largely driven by compensation) may represent over 75% of the costs in this people intensive business, and the rest is related to materials, equipment, and other infrastructure. A typical R&D program involves various specializations such as clinicians, toxicologists, engineers, chemists, biologists, statisticians, and others. The program also requires supporting entities such as information technology, infrastructure, accounting, and management. So, in addition to the costs of people involved directly in the production and testing of the prototype, we also have to include "indirect" costs related to the support entitles. Costs thus have a variable component as well as a fixed component. Fixed costs represent infrastructure costs and some of the support costs that cannot be avoided in the short run. R&D programs are also conducted in stages, so costs have to be estimated for each stage separately. In each stage, certain quantities of the NCE are manufactured. The drug substance is then formulated into a drug product (such as a tablet or capsule) that can be used in an experiment. In certain animal experiments or early human experiments, the drug substance could be used directly without formulation. In the manufacturing of the drug substance and subsequent formulation into drug product, one has to estimate the quantity needed for the current experiment and possible future experiments. Further, since the manufacturing and formulation processes have to be "invented," the precise yield (the amount of finished goods as a function of raw materials) is also not known. In making decisions regarding a specific R&D program, one should only include costs that are "avoidable" (or marginal). This is a function of the nature and timeline of the program, and since one cannot precisely predict the timeline or the complexity of manufacturing and testing beforehand (because of all the uncertainties discussed), one can get only an uncertain estimate of the costs of the proposed R&D program.

The second type of uncertainty is in timelines and project plans. The R&D programs in pharmaceuticals are discovery and development programs. They start with a hypothesis regarding how a disease can be treated, cured, or prevented by effecting certain actions in the body by the introduction of a chemical entity in a specific fashion. Since none of the components of this hypothesis are precisely known, the manufacturing of the drug substance and drug product for conducting the experiments requires the creation of a new manufacturing process. This process requires starting with raw materials that may be purchased from many vendors around the world or only a few specialized ones. Uncertain lead time exists between the order and arrival of starting materials from a vendor (who has to deal with other sets of uncertainties related to orders from many customers, availability of equipment and raw materials, shipping, and regulatory clearance). Once the starting materials arrive, further manufacturing delays may follow related to availability of personnel and equipment, availability of the specification of the manufacturing and quality control/testing processes, and the design and configuration of the manufacturing kit. Since the yield from each manufacturing step is not known, steps may need to be repeated to manufacture the necessary quantities. Since the impurities in the manufactured drug substance have to be precisely controlled (as impurities may have unanticipated toxic effects not related to the NCE itself), certain batches of the manufactured final drug substance or intermediates may have to be discarded if found to contain unknown impurities or known ones outside control limits. All of these possibilities make the timeline for the manufacturing of the drug substance and drug product to the required quantity uncertain. To make matters more complicated, there are compounding timeline uncertainties on the testing side. Experiments have to be run on animal and human models in stages. In animal models, availability of specific kinds of animals as well as personnel to conduct the experiments may introduce delays. In human experiments, selection and establishment of investigator sites (clinicians who conduct the tests), enrollment of subjects or patients (people who volunteer to undergo the study), and unanticipated and unique side effects (sometimes resulting in the shutting down of the experiment in a person, site, or the program) all can introduce uncertainties in the timeline.

Third, there are technical uncertainties in R&D programs. In each stage of the R&D program (and virtually at any time), new information arrives from experiments that may result in a reestimation of the anticipated success rate of the program. For obvious reasons, the R&D program will proceed only if the expected success rate is high enough and if the expected profits from manufacturing and marketing of the drug are high enough. I use "high enough" here as typical decision processes are based on "qualitative metrics" and not on a holistic shareholder value metric. Two types of "bad news" could force a "rethinking," "slow down," or even "abandonment" of the R&D program. The first one is the lowering of the therapeutic index, either due to lower than expected efficacy (the drug does not produce the beneficial effects anticipated) or a higher than expected toxicity (the drug has higher than expected toxic effects) or both. Such information arrives from the animal tests or clinical experiments conducted. The second type of bad news is that the cost of manufacturing the chemical is substantially more than expected due to higher than expected raw material costs, lower than expected yield, or higher than expected equipment costs. There can also be "good news" (the opposite of events described for bad news), which can result in a decision to "accelerate" the program by increasing the enrollment rate of patients and subjects by increased incentives. In either case, arrival of new information may force us to change our expectation of "success" of the R&D program.

Most often, the success rate is considered to be a binary outcome—it either succeeds or not. But, it is not always the case that the program suddenly dies. New information may force a redesign or modified expectations on the label (what can be sold in the market) and continuation of the program along a different development path.

Finally, there are uncertainties in the market potential of the product if it were to make it through the regulatory gates. Pharmaceutical R&D is a long-drawn process, often taking over a decade from idea to market. For many decision makers in R&D, this is a long time, and considering the market potential in R&D, decisions are often difficult. However, market potential is obviously an equally important component of R&D decisions. The expectations around market potential are unclear when the program starts and continues to change all through its progression, affected by competitive and regulatory actions, label expectations, therapeutic index, and a host of other factors.

In making a decision in R&D, all one can do is to capture all the uncertainties and use them systematically. The primary components of selection and portfolio decisions are, as discussed, costs, timelines, risks, and market potential. To improve decision quality, all available information has to be considered simultaneously. Often, different departments attempt to minimize or maximize components of the information structure. For example, manufacturing may try to minimize cost, project planners may try to minimize timelines, and marketing may try to maximize market potential with a more exhaustive label and adding in features for consumer convenience that allows a higher pricing premium. Independently, all of these make rational sense. However, they are not independent activities and improvement in one dimension may have adverse effects in others. The best decision can only be made if all interacting and interrelated uncertainties are considered simultaneously so as to maximize economic value. Existence of a complex set of uncertainties should not be the cause of adoption of ad-hoc decision processes. Shareholders do not pay decision makers to gamble; they expect them to improve decision quality. Decision quality is related to the systematic use of all available information. The information structure in a company thus should include not only the relevant information for decisions but all the uncertainties surrounding it.

Infrastructure

Now let's study the third component of structure—infrastructure. Infrastructure includes both the real infrastructure (buildings, plants, real estate) and financial infrastructure (capital structure) of the company. Typically, the real and financial structures of the company are managed by different departments in a disconnected fashion. The decision to deploy a rigid physical infrastructure for the company was virtually automatic in the era of the Industrial Revolution. Most organizations existed to make tangible things and that required plants to house the machines and other buildings to house the humans. The scale (size) of the operation was of concern to the managers of these companies as they found that there are many ancillary services that can be utilized more efficiently when they have a larger plant in a single location. Thus, when decisions were made to create physical infrastructure, they tended to make large buildings and plants in anticipation of future growth as well as to take advantage of lower unit cost of production. In many cases, the design of the infrastructure focused on the inputs (raw materials, people), the transformation process (machines), and the outputs (final goods). Bleak and large buildings were constructed with no windows in an effort to streamline the manufacturing process. As industrialized nations moved from goods manufacturing into services, they kept the same ideas in the design, construction, and maintenance of physical infrastructure. The managers and owners of modern companies continue to believe that the physical infrastructure defines their companies. As such, most companies have centralized operations where a large number of people come to work in a few large buildings. Since the presence of the worker was an important proxy for productivity in the Industrial Revolution, managers still believe that if the worker is present within the company's infrastructure, productivity is likely to be higher. Although this may have been true in companies that made standard products (nuts, bolts, cars, etc.), it does not follow if the company's primary business is the generation of information.

In contemporary companies, the headquarters and the operations are typically kept separate. Senior managers in the headquarters focus on the financial structure of the company while the managers of production facilities focus on the physical infrastructure. An office, the modern day equivalent of machines and plants for the services industry, still holds significant importance to companies today. The size, appearance, and location of the "corporate offices" are aspects of pride for the pyramidal leaders of typical companies. The "address" may have influence on how the customers and suppliers view the company. However, once the company succumbs to such practices and builds a large corporate office in the middle of a large city, it loses flexibility in the infrastructure dimension. To design flexibility into infrastructure, a company has to start from its objectives and business model. If the company is in the services business, it is unlikely that it has to own physical infrastructure for optimal performance. Ownership and maintenance of physical infrastructure is a business in itself and there may be companies specializing in this area. The separation of corporate offices and the operations of the company is an archaic notion created by the pyramidal leaders on the false belief that there is a separation between the financial structure (white collar) and physical structure (blue collar) of the company. Most believe that these are different jobs with no need for interactions between them.

Many information-related modern companies, such as pharmaceutical companies, have taken the same approach to marketing (corporate) and R&D (operations). Marketing and corporate finance reside in glass towers in the middle of a city while R&D may be in the suburbs, driven by scientists. They are considered separate and unique. The leaders also typically have different skills—the corporate office is driven by financiers/marketers/lawyers and the R&D operations structure driven by scientists. But if the company's objective is to create information (new drugs are a proxy for new information) and deliver it to biological systems to prevent and cure diseases in a fashion that allows the company to have a sustainable business, such rigid demarcations among finance, marketing, and R&D do not help it realize this objective. Segmented infrastructure is an idea that managers of industrialized companies implemented to have higher control over operations. For example, an automotive company may have different departments for the production of components of the automobile. This way, the outputs of each department can be measured and assigned a productivity score. The same ideas are incorporated into modern companies too by the segmentation of finance, marketing, and R&D. However, given that the outputs of these "departments" are not widgets that can be easily measured, a functional demarcation coupled with the practice of centralization in physical infrastructure leads to specialization in silos. The marketing personnel downtown are constantly thinking about how to price and sell existing and new products and since the marketers are surrounded by other marketers, most of the time is spent in just marketing, regardless of the current and future product portfolio. Meanwhile, in another building or on a separate floor, the financiers and accountants struggle with the financial matters of the company. Much time is spent on forecasting revenue, costs, and profits. Margins are analyzed and forecasted. Earnings per share and other accounting metrics are also forecasted and "communicated" to "the street." The buy side financial analysts gobble up the company's forecasts, debate, agree, disagree, and reforecast for the advantage of their clients. Much of this activity is conducted in a vacuum without much understanding of what is happening in operations and R&D. As the quarter draws near, forecasts are refined a number of times and finally when the much anticipated day arrives at the end of the quarter, the company's finance department hold their numbers up high, having met or exceeded the targets or in some cases lock themselves up in shame for having missed the estimates. This movie is played over and over again, and much of the management time is spent on managing the finances; this has long-term implications for the real value of the company.

To understand flexibility in the manufacturing infrastructure, consider the R&D manufacturing in a pharmaceutical company. Research and development manufacturing in pharmaceutical companies faces significant uncertainty. Manufacturing is typically divided into two areas: drug substance and drug product. Drug substance manufacturing deals with the manufacturing of the active pharmaceutical ingredient (API). Pharmaceutical companies can invest into creating API manufacturing capacity internally or outsource manufacturing to contract manufacturing organizations (CMOs). Creating internal capacity is costly as it often requires heavy up-front capital investment. The demand for API manufacturing (I will use batches as a unit of measure) is a function of the R&D pipeline dynamic of the pharmaceutical company. It is a function of the stock of candidates in the pipeline, the position of candidates in the R&D process, and the flow of new candidates into the pipeline. If R&D is creating candidates at a healthy rate or the stock of candidates in the pipeline is high, the demand for API batches will be high. On the other hand, if R&D productivity is slow, demand for API batches will be low. R&D is also a creative process that may not work like a manufacturing process that produces at a constant rate. Ideas arrive randomly, and ideas progress up the R&D pipeline with uncertain timelines and success rates. This means that the demand for API batches is uncertain in terms of both quantity and timing. It is also the case that pharmaceutical companies experience "bunching of candidates" in the pipeline for a variety of reasons that result in a bolus of candidates at one time and corresponding increase in API demand followed by periods of low activity.

Investing in internal capacity takes committed investments up front. It will create a fixed capacity internally. Because it is internal to the company, the marginal cost of production of an API batch will be lower, after the upfront investment is taken. Creation of a fixed internal capacity, however, reduces flexibility for the company. If demand is low, the up-front investment expended in creating fixed capacity will be wasted. A strategy of outsourcing API manufacturing will allow the company to match demand with capacity on an as-needed basis. It, however, comes at a cost. The suppliers, CMOs, will demand premium pricing, and the marginal cost per batch of acquiring API from vendors will be higher than internal production. CMOs have made an investment in capacity on the basis of future demand and will require a return on that investment. Since they can manufacture for many different customers, they have a higher level of flexibility in managing demand and capacity compared to a single pharmaceutical company. Thus, it is possible that the cost of acquiring an API batch from outside is lower than internal cost. More importantly than cost, however, is the flexibility the company gains to switch production between internal and external sources to better manage future uncertainty. Many companies are using outsourcing as an important flexible lever in managing and meeting demand. High internal capacity is costly when the demand is variable as it reduces flexibility in demand and capacity management.

Increasingly, these outsourcing contracts have become an important strategic consideration for manufacturing functions as they have to consider many aspects such as the risk of not meeting demand—internal and external costs (including marginal and total costs), cost of excess capacity, management, and logistical complications of outsourcing and the associated costs and economic gains. Often, such contracts are entered on an as-needed basis (ad-hoc and tactical sourcing) or on a template for long-term contracts (with rigid contractual obligations). Because outsourcing is a procurement function, it tends to have characteristics of arm's-length transactions and thus eliminates opportunities for both supplier and buyer to introduce flexibility and thus enhance value through better management of uncertainty. Principles of procurement and negotiations currently practiced by large companies are remnants of the Industrial Revolution and are not very useful in a world driven by uncertainty and the need for flexibility.

There are a number of features that both the supplier and buyer can consider in contracts, such as:

Option to terminate
Option to delay
Option to accelerate
Option to increase or decrease committed quantity
Option to change delivery schedules
Option to prebuy certain quantity or capacity
Option to abandon sequestered capacity
Option to cancel or change orders

Outsourcing contracts can act as a bridge between the pharmaceutical company's physical infrastructure and its capital structure. By integrating the two, companies can gain a higher level of flexibility as it has a higher number of dimensions to influence as uncertainty resolves in the future. These contracts are also a big component of risk management in these companies. By designing them holistically with embedded flexibility through various features and options, the company can reduce overall risk. If the suppliers are located in various geographies domestically it may also reduce loss of service risk. If the suppliers are international, it may allow the company to match its currency exposure on the revenue side with exposure on the cost side. If the company is actively pursuing to hedge currency risk in revenue, international suppliers may allow lower or eliminate hedging, saving the cost of hedging as well.

Creating infrastructure in a modular fashion will also enhance flexibility. Imagine a manufacturing plant with a certain capacity producing some end product, which can be a chemical (as in life sciences or commodity chemicals), an electronic component (as in technology), finished materials of a certain grade (as in mining), gasoline (as in a refinery), or any finished good in any industry. The capacity may be a function of equipment, people, space (for manufacturing and storage of raw materials, in-process inventory, and finished goods), and availability of manufacturing components (e.g., raw materials, additives, electricity, water). The capacity will require some lead time before it can be functional for the production of the finished goods. Managers of supply chains and manufacturing and demand/capacity balancing struggle to ensure that their capacity utilization is high enough for a profitable operation but not so tight it creates a delay in the system. Often, the lead times involved in creating capacity are long, so managers have to anticipate future demand well in advance and invest in creating optimum capacity for it. If they over-design the supply chain, the result may be excess and idle capacity (increasing costs). If they under-design, they may be forced to find alternative suppliers for excess demand (which cannot be met with existing capacity) or accept a delay in delivery (and related loss of revenue, increase in costs, or both). If the equipment can be modular, each module representing a small capacity, the manager can match demand to capacity by selecting the level of modularity (the capacity of equipment) that can be incorporated into the design. Once the modular equipment is available, the manager can select the number of modules needed to be put into operations (perhaps each year) by observing demand. The flexibility to closely match capacity to demand can add tangible value to the company.

Let's run through an example to demonstrate the power of modularity. Consider a manufacturing plant with production equipment designed to be modular. Each machine has a small capacity and there are number of such machines available. There is a cost to operating the machine but if the machine is left idle, no cost is incurred. Each year the manager will decide how many machines to activate after observing the demand for that year. There are no costs to activate or mothball machines. The manager's decision each year is an option since it gives the flexibility (the right but not an obligation) to select an optimal number of modules to meet the demand. Every year, the manager selects just enough machines to meet demand but not exceed it. For example, if the demand is 200 and the capacity of each machine is 15, the manager will assemble the plant with (200/15) = 13 machines. Any remaining demand is outsourced to an external supplier. This ability to closely match demand to capacity through modularity can increase the value of the company.

Companies may design infrastructure in a distributed fashion to increase overall flexibility. Consider the case of a technology company that brought a new consumer electronics product to market, a wireless phone that can be used anywhere in the world and can make calls over the Internet as well as on a regular phone line. Demand has been strong in test markets, and the executives at the company have been discussing better ways to organize the supply chain so that they can manage demand across the globe. For a new product like this one, it is often difficult to predict demand in various geographies. Because the base unit of the phone is plugged into a power outlet as well as a phone line, it has to have different configurations for different countries. Countries have different power outlets, voltages, and phone jacks. This has caused a major headache for the manufacturing and logistics staff. Marketing has been having problems projecting demand. Both the newness of the product and the various price points and geographies the product could be sold have made it difficult to do this. The company could project an expected range of demand, but pinning it down to more precise estimates is nearly impossible this early in product introduction. Manufacturing department's problem was that if they manufacture the product for one country (say, Japan), it will not be usable in another (say, India) as the countries differ in power and phone infrastructure. In essence, the company must manufacture a customized product for every country. If the product takes off in one geography and not in others, the company cannot switch shipments because of this customization. However, if the design of the new product is such that the ultimate assembly can be done closer to the consumer and later in the supply chain, the company can keep its options open. For example, if it ships generic units to a central warehouse in Asia and the customization components to the various country depots, it can manage uncertain demand better. If demand really takes off in India, the company can send the necessary base units to the local depot in that country and customize them for the Indian market. On the other hand, if demand is high in Japan, the company could divert the base units to that country, combine demand in Asia for the base units, and manage the production for the total demand. Such a flexible infrastructure will allow the company to delay customization, manage uncertainty better, and increase value.

The supply chain is an important strategic issue for manufacturing and logistics companies. Many companies attempt to reduce uncertainty for improved performance of the supply chain. Although information that allows better prediction (and thus lower uncertainty) is good in the management of complex supply chains, the economic value gained from introducing flexibility should not be discounted. Companies such as Wal-Mart and Hewlett-Packard Company have long recognized this and have invested in increasing flexibility. In many cases, the success and failure of companies may depend critically on how they design manufacturing and logistical systems as well as the features they incorporate into supplier and buyer contracts. Also, it is not only the functionality of products that is critical but also their industrial design. Ability to delay decisions and scale up through modularity are critical attributes of industrial design.

The ability to switch production is another aspect of infrastructure flexibility. For example, consider an automotive manufacturing company designing a new assembly line. The company is a large manufacturer of passenger cars and has been having a difficult time lately. Gas prices have been volatile, and consumer preferences for automobiles are more unpredictable every day. The design department has been busy putting the final touches on two new models: a sports model called the Zephyr and a hybrid intermediate called the Electra. Both are revolutionary designs and will require building new assembly plants as both the body and the components are of new materials, and these models are manufactured very differently from existing models. The design department has been careful to design them to share certain components and materials as they have been very aware of the unpredictability in the market both for demand and preferences for types of cars. For the new manufacturing plant, the company is considering two different designs. Both designs have two assembly lines. In design A (called fixed), each assembly line is specialized. That is, each line can only make one type of car. Line 1 will make the Zephyr, and Line 2 will make the Electra. In design B (called flexible), assembly lines are not specialized, and both Line 1 and 2 can make either car model. The flexible design is more expensive and will require a higher level of investment than the fixed design. All previous manufacturing assembly lines at the company have been versions of the fixed design. All lines are specialized and can make only one type of model. The biggest advantage of the fixed design is cost. It is cheaper and easier to design. Given the cost advantage, some members of the executive committee questioned the consideration of the flexible plant design as they considered it just a waste of time, money, and space. They feel that although the future is unpredictable, the flexibility offered by the other design is unlikely to justify the cost. The debate has accelerated as a small cohort of young plant managers started pushing for flexible designs. They feel that it does not make sense to have specialized lines for single models as the demand for them is impossible to predict. The utilization of specialized lines will be less, they argued, and that alone may justify the additional investment. Who is right here? Intuition tells us that the flexible design is likely much more valuable given the high uncertainty in consumer preferences, macro-economy, and gas prices. However, the decision depends both on the cost of construction and the future benefits. Only an analysis that considers uncertainty in product demand, construction timelines, costs, and risks can answer which design is better in this case.

Creating or implementing the infrastructure in stages is another way to increase flexibility. Consider a manufacturing company thinking of implementing a new information technology (IT) system. The executives have been debating over the implementation of a new IT system to improve the productivity of employees for several months. It all started when some of them attended an IT conference where a large software company exhibited its productivity enhancement system (PES) to improve workplace productivity. The company showcased ongoing implementation at several of the company's competitors. This worried the chief information officer (CIO) of the company, who felt that if the competitors implemented the system, it may be important for the company to take the plunge as well. The CFO was against it and remained convinced that the massive up-front investment to implement the technology was a huge risk as they did not have much data to show that PES in fact improved productivity. At the conference, a PES executive described how important it is for companies to implement the system across the entire enterprise. The executive showed that an enterprise-wide implementation lowered licensing fees (on a per seat basis) as the company would be purchasing a large number of seats together. PES pricing was a function of how many copies the client purchased in one implementation. PES had scale advantages in terms of people and equipment, and the pricing included implementation. The executive showed statistics that indicated that many of the PES clients decided to implement the system across their entire companies to take advantage of the favorable pricing. "I am not really sure," said the CFO. "How do we know if PES is effective and how much productivity improvement can we expect? We have 100,000 employees and five different departments across a dozen countries, and it will be a big investment for us up front. If the expectations are not realized, it can have an adverse impact on our stock price." The CFO also felt that the fact that the competition was implementing PES was not sufficient for the company to take the plunge. In a recent management meeting with the CIO, one of the technology managers suggested that perhaps it may be better for the company to try a staged approach to this implementation. In a staged implementation, the company will implement the system in one department or location, get information on how well it performs, and then make a decision to roll out further implementation to other departments and locations. This staged approach adds flexibility to the company and increases its economic value. Staging is an important component of flexibility in many areas such as R&D, manufacturing, and market launches.

There are other ways companies can add flexibility. For example, ability to select equipment from a variety of available equipment types is another important component of manufacturing flexibility. Consider an electricity producer. Electricity is a special commodity as it is needed for virtually every aspect of life in the modern world, but it cannot be stored effectively and at scale. Electricity has to be produced when it is needed. If the supply of electricity is not commensurate with demand, the price will spike as the marginal user will pay the market-clearing price. Since total demand cannot be precisely predicted most of the time, demand and supply imbalances and corresponding price fluctuations do occur. Electricity producers deal with the problem by having a network of plants, some producing power constantly (base load plants) and some operating only when there is a spike in demand (peaking plants). By optimizing production between the base load and peaking plants, generating companies can closely match the demand. Base plants typically cost less per unit of power and they operate for long periods of time before a planned or unplanned outage occurs. Nuclear power plants fall into this category. Peaking plants are flexible and can be switched on and off as needed but cost more to construct and maintain per unit of electricity produced. They are typically fossil power plants that use fuels such as natural gas or oil. Both having a mixed type of plants in the network, base, and peaking as well as having the ability to use different fuels increase overall production flexibility for the company and increase value. The ability to select from production systems of differing characteristics can improve flexibility in many areas including manufacturing and logistics.

The financial structure of the company (another component of the infrastructure) is intimately related to the physical (real) infrastructure. The separation of the real and financial structure, as conducted in most companies has implications for flexibility and risk management. Financial structure includes the capital structure of the company, its ability to weather short-term storms (such as its cash position), the contracts it has signed with buyers and suppliers, the hedging and insurance contracts that may exist, and other attributes. For example, a biotechnology company with real assets in its R&D pipeline (such as R&D programs and technologies) has a certain level of risk emanating from it. To manage the overall risk of the company, it has to combine the real risk with its financial risk. R&D programs house uncertainty in many aspects and may experience high negative shocks at certain times. This may be due to the technical failure of a promising R&D program, delay in a program due to unavailability of materials, a patent infringement legal case, regulatory rejection, failure of a marketed product due to unanticipated problems, or a variety of other issues. A company with little financial flexibility will "break" under such shocks even though it still has a valid business proposition. If the company is capitalized with venture capitalists with rigid horizons and return expectations, it may be unable to weather setbacks in its R&D programs. So the flexibility in the capital structure includes tactical flexibility such as cash, long-term flexibility such as low leverage (or debt), and the flexibility allowed by the capital provider in terms of horizons, tactical returns, and objectives. The flexibility related to the orientation and objectives of the capital provider is clear in private markets but less so in public markets, where a large number of investors with varying objectives and risk/return trade-offs exist. Managers of the public companies tend to fault Wall Street and investors for their short-term focus and lack of understanding of the company's value. However, when there are a large number of diverse investors, the loss of flexibility due to misalignment of investor objectives with that of the company are minimal. The perception of public investors having a short-term horizon is created by managers whose bonuses and employee stock options work in defined windows and so the loss in stock price in a short-term window has an impact on the total compensation of managers. The criticism of Wall Street for short-term blips in stock price is a way to divert attention from the real issue managers are worried about—the hit to their own compensation. Well-managed companies seldom worry about short-term movement in stock prices and do not "criticize" investors for not pricing their stock properly. The stock market, with a diverse set of investors, behaves reasonably efficiently.

The interactions of the real infrastructure and financial structure do not get much attention in companies. This is of great significance for holistic risk management and the maintenance of the right level of flexibility. For example, an electric utility that has physical infrastructure for the production of electricity (generation assets) may have financial contracts in place for fuel suppliers and electricity buyers. The terms in these contracts may have a variety of features including options to buy more or less, cancel, graduated price breaks, and other factors. The company may also maintain multiple contracts for the procurement of electricity from other providers to manage outages in their own facilities. How the company utilizes the electricity plants (physical infrastructure) is intimately connected with the existence and features of the financial contracts they have with customers, suppliers, and partners. Some of these contracts may be short term such as hedging contracts bought and sold in open markets and some may be long term such as forward contracts for the procurement of fuel and sale of electricity. Such contracts may be initiated and managed by the finance function within the company while the operating decisions may be made by other departments. The inability of such companies to manage overall risk (real and financial) in a complete framework drains flexibility and makes them vulnerable to two different types of shocks. A real shock such as an outage of a plant will be met with operating actions disconnected from the financial side. A financial shock such as a sudden increase in fuel prices may be dealt purely on the financial side without consideration to real flexibility that may exist. Disconnected management of the real infrastructure and financial structure thus introduces a level of rigidity in the company and commensurate loss of value.

Ad-hoc tax policies coupled with the company's desire to enhance short-term revenues and profits, perhaps to appease the investors, can decrease its financial flexibility. For example, if the company has tax incentives to shift production abroad and/or keep profits abroad, it may do so in the short run with an associated bump in after-tax profits. However, in doing so, the company has not eliminated the tax liability, just shifted it to a later time. The company's balance sheet may portray short-term flexibility, but it may be cosmetic, as the cash in the balance sheet cannot be accessed easily. Since repatriation of such cash will incur a tax penalty, the company may forgo the use of its own cash and may raise debt in public markets. Thus, lack of access to cash (and short-term flexibility) may force the company to lose long-term flexibility by the increase in leverage.

In summary, flexibility in the human, information, real, and financial structure is critical for the success of the modern organizations dealing with uncertainty. Contemporary companies are ill-equipped because of the status quo ideas of management primarily designed for the Industrial Revolution. Many companies may have to essentially start over by systematically introducing flexibility in these dimensions for future success.

5

System

There are three important systems in organizations that need to be designed with flexibility as a fundamental driver. They are technology, process, and content.

Technology Systems

Technology systems include all application of technology in the enterprise including but not limited to information, communications, manufacturing, and building technologies. Today's companies consider technology systems in segments and manage them in different departments. *Information technology* (IT) has come to represent technologies dealing with the collection, storage, and reporting of data, primarily driven by computers. *Communication technologies* typically represent telephony and related aspects. *Manufacturing technologies* imply robotics and automation. *Building technologies* represent the creation and maintenance of buildings, plants, and transportation structures. Although these technologies share many common and fundamental features, they are designed and managed based on the end use rather than treating them as a system that touches all aspects of an organization.

From the Industrial Revolution, application of technology has been largely to reduce costs, improve timelines, and enhance efficiency. Technology was considered to be a substitute for labor. When labor costs were high, products with higher technology content did better. To improve the technology content in products and services, companies sought automation. As power sources that can be used to automate activities grew—from steam to electricity—a wider selection of automation technologies became available. Switching between labor and technology in production processes was a real option (flexible decision) held by companies in that era.

For the owners and managers of the companies in the industrial era, technology came with added perks—no absenteeism (except occasional breakdowns) and consistency in productivity (items produced per unit time). Labor was notoriously fickle in this dimension in spite of the watchful eyes of the circulating supervisors and managers. For the companies in the modern information era, technology continues to be a mechanism to reduce the labor content in their products and services. However, low-cost emerging countries with artificially low currency rates (by government fiat) have provided a temporary relief mechanism in this score. In the last several years, many service companies moved away from increasing the technology content in their products and services because of the low labor costs available in emerging countries. *Outsourcing*—a euphemism for maintaining the labor content in the products and services by shifting labor to low-cost countries and delaying adoption of automation technologies—has been fashionable in many industries. For the managers of these companies, outsourcing was a quick way to show tactical profits, but many did not realize that they are losing flexibility by doing so.

Many emerging and low-cost countries, buoyed by the phenomena of outsourcing, directed much of their export economy into sectors where they had an advantage—continuously increasing the supply of low-cost labor. This has helped the companies in the developed countries to slow down technology adoption and continue to take advantage of the low labor costs. Emerging countries accomplished this outcome through policies of an artificial exchange rate and concentration of efforts in a few export sectors. In the long run, emerging countries may have set themselves up for a significant negative shock when foreign companies resume increasing technology content or the pressure on the current artificially low exchange rate becomes unsustainable.

Decisions to adopt technologies in companies to improve productivity and profitability are important from a shareholder value perspective. Often, such decisions have a critical impact on the viability and success of a company in a dynamic and hypercompetitive world. Technology has a big impact in the design of facilities and manufacturing plants as well as the management of logistics. It also affects how the company innovates, designs its products and services, and manages the life cycle.

In the 1990s, many companies invested massively in information technology (IT)—in the collection, aggregation, storage, and reporting of data as in enterprise resource planning (ERP) systems and activity tracking systems (ATS). Investments were aided by the promise of "huge" productivity improvements by software firms that created the technology and firms that assisted in implementing them. The jury is still out on whether such investments actually helped the companies. Software companies calculated a return on investment (ROI) for the implementation of their suggestions to convince their clients of the value. Such ROI calculations were made using fixed expectations of costs, timelines, and benefits. As anyone involved in such implementations knows, costs, timelines, and benefits could not have been predicted with any level of certainty. The risk of the technology implementation not delivering expectations was also ignored. In many cases software companies were successful in convincing the companies that they should implement the new systems in a single stage as such implementations would have reduced costs. In all these cases, technology implementation was pursued for the sake of adopting a new technology without really understanding how that affected flexibility.

In considering technology and design improvements, it is also important to understand the impact on the entire system. In manufacturing, for example, an improvement in some aspects of the supply chain (such as reducing internal capacity to save costs) may have deleterious effects in other aspects (such as stock outs). Supply chain optimization techniques have been applied in many cases to improve performance, but the metrics used to measure improvement (e.g., cost, speed, availability) do not automatically imply an improvement in either flexibility or economic value. Although reducing cost, increasing speed, and increasing availability (reducing stock outs) could be beneficial, it is impossible to say that taking actions to optimize these metrics individually has any beneficial effect on the entire system. The influences of the systems utilized by a company are not just contained within the company but extend to its customers and suppliers. How the company designs the interactions among its suppliers and customers has an important bearing on the performance of the supply chain. Supply chain managers typically focus on reducing uncertainty rather than increasing flexibility to manage it.

In the modern information companies, information technology (IT) has taken a special significance. Software companies have been creating bigger and more elaborate programs in an effort to increase efficiency in the enterprise. In this context, IT is very similar to steam power from an earlier age. Both are used primarily for automation. In the case of IT, there are many dimensions of automation such as collection, storage, and reporting of data. Since information is critical for modern companies, data collection, storage, and reporting are considered to be important functions by the leaders of traditional companies. Since different companies provide the hardware and the software as well as the implementation services, companies who have been attempting to solve similar problems have ended up with very different outcomes, both across their own organizations as well as across the industry.

There are three fundamental problems that introduce rigidity into the design and use of information technology in companies today. First, information technology implementations are disconnected within departments and other technology systems in the company. This diminishes the ability of the company to use it effectively, react to new information fast, and apply it across all aspects of the business. Second, information technology focuses on data, much of which may not have any use for the company. Since "data" is always considered to be a "good," all available data are collected and stored using increasingly cheaper systems and storage devices. This increases confusion and reduces the company's ability to make decisions under uncertainty. Existence of data may spawn many different analyses in the company to aid decisions, based on the belief that more data and analysis are always good. This creates long and complex decision processes, reducing the company's ability to react quickly to new information. And finally, contemporary IT implementations are long and time consuming processes. Both the providers of the equipment and software as well as the firms involved in implementing them may have incentives to create complex and long implementation processes. For example, a complex system increases switching costs for the company. Since most IT systems have ongoing maintenance revenue for the providers and implementers, they have the incentive to ensure that the company continues with it for a long time. However, a long and complex implementation process makes the end product less relevant by the time it is functional, as technology may already have changed and the "data" gathered and reported by the system may be irrelevant in the company's current business model. The switching costs added to the system further reduce flexibility for the company to move to more relevant technologies quickly.

Applications in the information technology realm often attempt to automate an action—data cleansing, analysis, reporting, and so on. From the early years of computers, the stand-alone nature of the machine prompted stand-alone applications. Stand-alone applications attempted to improve the productivity of a single user. This is similar to the tools used to make certain manual tasks easier in the industrial companies. Early applications—such as those for word processing, spreadsheets, and presentations—resulted in a loss of productivity for the individual user. This is because, unlike a manual tool that automates a defined action in manufacturing, application programs create nonspecific outputs. Word processing, for example, does not have a defined objective unlike a metal bending tool that is used to bend a specific part in the production process. Since the word processing program can be used for a variety of actions and to create a variety of outputs, how the individual user makes use of it is critical. For example, if the individual user could have written the document by hand on a piece of paper and communicated it faster to the user of that information, the use of the word processing program for the same purpose actually results in a loss of productivity. This may happen if the user spends time formatting, such as changing fonts and margins that may not have any effect on the end outcome. Productivity increased nearly a decade after such programs were introduced as it became possible to share information electronically, over the Internet. Most of the productivity gain from application programs thus came from not because of their stand-alone use but the ability of the user to share information with others electronically.

Sharing of information, thus, is an important ingredient of technology systems. Because of the rapid increase in the availability of data, teasing out relevant information from mountains of available data has become a tricky business. Since our ability to internalize data is limited, most of us have to select a small manageable amount from a large available set. Since the utility of data is situation specific, it is difficult to predict the most useful data in a generic fashion that is applicable to everybody and to every situation. Thus technology systems have to incorporate the need for delivering the right information at the right time and not just the collection, storage, and reporting of data that does not have a context.

Consider a technology system that is designed to encompass information, communication, manufacturing, and other systems such as buildings and transportation with a focus on decisions. Information that is relevant for decisions has to include uncertainty and interactions. For example, the decision to manufacture certain units of one of the company's products may depend on internal capacity available, projected uncertain demand and pricing, and transportation and logistics considerations as well as the possible effects on other products and services offered by the company and competitive and regulatory response. If manufacturing is automated, decisions will have to be made within the plant by robots or similar systems that need to also internalize all available information. Today's organizations may have reached a point of complexity—due to accelerating uncertainty and interactions—and machines may for the first time be positioned to make better decisions than the managers.

Enterprise resource planning (ERP), enterprise resource management (ERM), customer resource management (CRM), and a variety of such systems have provided a perception of holistic consideration of available information. However, these systems provide modules that are specific to different parts of the business. Finance and accounting, manufacturing, marketing, and others all have different modules for the collection, storage, and reporting of segmented data that meet the tactical requirements of managers. In designing technology systems, an organization needs to consider all decisions and interactions (it should encompass the entire organization), the type and timing of information and actions to be taken (uncertain information and actions based on them), and how to deliver and internalize such information to be effective.

Another important aspect of system design is the flexibility associated with platform technologies. A platform technology is one that allows a company to create expansion options in the future and this adds overall flexibility. For example, a manufacturing company considering automation can design systems in different ways. It can design and implement a system that is custom tailored for specific automation tasks that exist today. Such a design will be cost effective and more efficient as it is made to solve the problem at hand. However, if the product designs change and require different machines, the automation technologies have to be replaced as well. Alternately, it can design a platform automation technology that can be applied in many different situations. It may be more expensive to design and it may be less efficient to solve the current problem. However, it provides the company with a platform that allows modifications as needs change in the future. It also allows the company to possibly create newer automation techniques in the future, providing a higher level of flexibility.

For life sciences companies, platform technologies add important flexibility. Consider a biotechnology company with a platform technology that allows it to modify existing molecules to make them more effective. Such a technology will provide the company with flexibility as the company can apply it across disease areas and phases of research and development. Such a company has less technical risk compared to another company that is pursuing the development of specific molecules for targeted diseases. R&D programs in specific molecules or mechanisms of action have program specific risk. A company pursuing a limited number of such programs runs the risk of catastrophic failure. Platform technologies thus may have a significant influence on the performance of the company's portfolio and its eventual success. Creating an environment where the platform technology can be broadly applied will help the company take maximum advantage of it.

Technology also includes systems for compliance to explicit and implicit regulations. For example, the financial system can be viewed as one that not only allows the company to ensure compliance to explicit financial regulations but also provides consistency in operations, measurement, and incentives. Typically, large companies have a myriad of financial systems, some specializing in external reporting and some others focused on internal operations. They are kept separate and information from one is transferred to the other manually. Many companies view the external financial system as a "compliance system," something they have to do to appease the regulators and investors. As such, the information in such systems has to be closely managed so as to be consistent with the company's projections and the market's expectations. Many companies also believe that they have to manage the markets by providing explicit forecasts and meeting or exceeding those forecasts. Thus, financial systems, focused externally, are designed to forecast and to meet the forecasts. Meanwhile, the internal financial systems are designed to measure and monitor inventories, manufacturing capacities, costs, and other aspects. The entire financial system of the company should ideally be one that integrates external and internal activities. The self-imposed constraints by corporate finance to "precisely forecast" the company's earnings and then meet those forecasts through a variety of accounting and real actions reduce flexibility in the financial systems in the company. If corporate finance managers can be taught that the markets are smart in understanding the company's future prospects (regardless of the accounting and strategy stories that may be fed to them), such futile activities can be eliminated and flexibility restored to the financial system. By connecting the external financial systems (focused on financial assets) and internal financial systems (focused on real assets), the company can introduce flexibility in overall management of finances. In such a case, the investors of the company will have transparency into the company's operations and do not have to be provided forecasts and stories about the company's future.

Process Systems

A process is a set of actions that may be repeated so as to obtain an expected and routine outcome. Humans are less adept at executing processes compared to machines. This is because machines are less prone to emotional and irrational decisions and differing actions in the presence of the same information. However, from the pyramid makers to the Industrial Revolution to today's companies, humans are largely deployed in executing processes. In the early days of humans, life had been much more challenging and unpredictable. Although the end objectives of food, water, and shelter were well defined, the process by which each individual would have met these objectives was largely undefined. There was very little repeatability when and where the next animal can be hunted and where danger remained hidden. Humans had to be creative and use their brains to survive and succeed in such an unpredictable world. However, in the last 5,000 years, high repeatability of activities emerged, aided by well-defined end outcomes and processes to reach them. In this process-centric world, humans are delegated to repeating routine activities set to specifications. A premium emerged for those who can repeat a specified set of activities with less error in the world of pyramid makers and industrial companies. This transformation of humans from free-wheeling hunters and gatherers to operators of routine activities took place in a relatively small period of time (a few thousand years). Humans evolved to dominate their environment primarily because of their brains and associated creativity. The delegation of creativity to routine repeatability led to the application of a highly evolved human brain to suboptimal uses.

The process-centric view of the organization remains dominant in today's companies. A large percentage of employees in a company still execute processes that have a repeatable set of actions with a defined sequence, timing, and specifications. This is true from the manufacturing floor to office towers. In most cases, performance is measured by the ability to conduct actions according to specifications, repeat without noise, and produce outcomes without uncertainty. There are incentives to define the processes better and ensure that they are static (a change in a process will require retraining and associated loss of time and increased cost). Performance can be measured by the lack of variability in the output of a process. What companies are forgetting in this mad rush to increase efficiencies in processes is that humans are thoroughly unequipped to conduct them. We are designed to roam the wild and use our brains.

One would think that the advent of computers and automation may have given most companies a sudden relief from stuffing incompetent humans in processes and replace them with reliable machines with no emotions. However, companies have been very slow in adopting technologies and automation in processes either due to cost, due to fear of the unknown, or just inertia in a change from status quo. With the short decision and performance horizons of managers, companies have built-in risk aversion to change. "Why fix something if it is not broken?" is the classical question that managers ask. So, if people are working well in a repeatable process why take the risk of introducing automation? More importantly, if automation is successful, what will the people do? The latter question was not asked by the benevolent managers but the workers themselves. Workers resisted automation because of the fear of loss of jobs. The cautionary tale of robots taking away all the jobs have been passed down from the Industrial Revolution to today. In the services industry, the computer is the "villain" that will make humans obsolete in the execution of routine processes. To be fair, the emotional, irrational, and unpredictable human is no match to a computer without such baggage.

The idea of machines replacing humans in repeatable processes presents a catch-22 for the society. On one hand, such a move will release human potential, currently tied up in performing suboptimal activities, and that can be deployed in more creative ventures. On the other hand, if such human capital is unwilling to go into more creative aspects, there will be an excess availability of humans who have nothing to do. Automation is understood to be synonymous with the reduction of the labor content in the products and services of a company. The workers will be released by the company if processes are automated. Thus workers always reject such a scenario. Managers also consider automation as a replacement to labor and for them this makes sense if the cost of production can be lowered.

But there is an alternative. For example, introduction of automation in repeating processes will provide the company with employees with higher levels of flexibility in their time and thought processes. Such an outcome will help the company innovate at a faster rate. This is often flatly rejected by the pyramidal leaders of large companies. Many believe that creativity is seen only in special places or it requires certain education and experience. This class mentality, that segregates human resources into buckets, forces the company to remain in status quo as it will always believe that automation and subsequent release of human capital has no beneficial impact on the rate of innovation. The economic value gained by the flexibility in workers' time and thoughts is often discounted and the decision solely depends on the investment needed and the payback period due to the lower cost of production.

As repeatable processes drive most of today's companies, many have embarked on documenting static processes to great detail. These process maps help companies to train new employees and identify performance issues. Early on, as companies introduced elaborate supply chains in their businesses, the managers were keen on utilization. In general, the feeling was that any break in activity is value lost and any excess capacity (labor and machine) is suboptimal. The attempts at running supply chains to very high utilization produced counterintuitive results for the pyramid and industrial machine makers. Full utilization of people and machines in a process results in a complete loss of flexibility. Any unpredictable event such as a machine breakdown or small levels of absenteeism brought the entire operation to a standstill, as there was no excess capacity anywhere in the process to temporarily replace the absent part. The importance of flexibility introduced by some level of excess capacity was fast understood by companies. As they widened the supply chains beyond their own walls to include buyers and suppliers, many sought to optimize the process by introducing optimal excess capacity at the right places and at the right time.

However, a well-oiled and optimized process imposes higher order rigidity on the company. If the process is static, the company will be unable to change it as uncertain future information becomes available. For example, a manufacturing company, with carefully selected warehouse capacities at optimal locations to serve the status quo demand of a singular product, may be rendered completely incompetent if the demand pattern shifts or consumer demand changes. Although the operational flexibility is important to better manage status quo, the design of the process should also allow fast adaptation to future changes. A high level of precision in the documentation of existing processes may make companies prisoners to their own static processes.

Technology has made a tangible contribution to the "processification" of the modern companies. Resource planning and management systems and manufacturing and logistics systems have reinforced status quo processes and attempted to extract efficiency by rendering them static. The net result of such technology implementations that better defines and executes status quo processes is a loss of process flexibility for the company. A flexible process is one that is able to adapt to new information. In a static process, the activities and the connections and interrelationships between activities are prescribed and maintained. For example, a newspaper company may have defined itself as an organization that provides news on paper and may have established efficient processes for the collection of news, printing of the newspaper, and delivery of it to customers. To extract efficiencies from this well-defined process, it may have designed a static format for the paper, a printing method and technology, marketing and logistics, and other aspects. As the company attempts to extract efficiency from the newspaper product by a tighter control on all component activities, thus reducing noise in the end output, it also loses flexibility. As many newspaper companies in the United States recently realized, they have been extremely efficient single product companies but people switched to getting their news in a different way, making their end product obsolete. The focus on extracting efficiencies from status quo processes left these companies in the dust, unable to adapt to the new reality. In retrospect, some may be seeing the importance of having the ability to switch at the expense of a less defined process and associated higher per unit cost. It may be good to have low cost production processes but if that is at the expense of the ability to adapt to new information, it ultimately leads to failure.

Today's airlines may be focusing too much on reducing the cost and increasing the revenue per mile flown and designing processes to do so. But if this is done in a vacuum, without understanding how customer needs and preferences for long distance travel are changing, the most efficient airline may be the first one going out of business. In a desperate attempt to decrease cost, some airlines have been reducing services also. If such a change accelerates consumers switching to alternative modes of transportation, the decrease in operating costs and tactically higher profits will come at the expense of lack of future profits. If consumers tend to see air travel akin to a trip to the dentist, it can lead to a long-term decline in demand for air travel. Again, the fine-tuning of status quo processes for immediate profits may spell trouble for the industry.

The pyramid makers may have created finely tuned and exceptionally efficient processes for the creation of pyramids. All activities—quarrying stones, transporting them, lifting them into position, and attaching them to the structure—have been done with exceptional skill and efficiency. This allowed them to create more and bigger pyramids as they deployed most of the resources in the economy to do just that. The focus on the process led them astray and not many questioned the need and utility of the end outcome. Automobile manufacturers may be facing a similar dilemma in the modern economy. In the last 20 years, improvements have been made in the process of manufacturing, assembling, and logistics of automobile parts and finished goods. Costs have been going down and the throughputs of the manufacturing plants have been increasing. All around the world many new automobile companies have sprung up, some more efficient in the supply chain and manufacturing processes than their more established competitors. The quality of the automobiles also has been increasing, leading to a longer life span for each produced vehicle. In the mad rush to optimize processes, reduce cost, and increase throughput, many have missed a larger question. Does the world need as many of these conventional automobiles with an antiquated engine that burns hydrocarbons and pollutes the environment? If the stock of operating automobiles in the world is declining at a slower rate (because of lower use and longer life span), what does that imply for future demand? Is the emergence of large consumer blocks in Brazil, Russia, India, China (BRIC countries) enough to replace the rapidly declining demand in other areas? Is the need for physical transportation of humans—to work, school, and shopping—declining with the advent of new technologies to collaborate, work, study, and shop online? Does it make sense to continue to increase throughput in the face of all these changes? If so, where should the production be and what type of automobiles? It is ironic that many automobile companies are teetering on the edge of bankruptcy as they proudly point to all the process efficiencies they gained in their manufacturing processes. As a senior executive of a failing automobile company famously said, "We make the best cars in the world. It is just that not many want to buy them." This is a sobering reminder to companies focusing on improving processes while forgetting the product.

The finely tuned and executed business processes in the financial world may have succumbed to a similar fate. For example, before the recent financial meltdown, large financial institutions were extremely efficient in the purchase, securitization, and sale of risk through complex and well-planned processes. These processes were architected with high efficiency so that they can scale with leverage. For example, once these companies created efficient processes for the purchase and accumulation of mortgage backed securities (MBS), they only had to increase the leverage to increase "throughput" of the process. Very much like a manufacturing company churning out products that nobody wanted to buy, at an ever faster rate, these financial institutions kept buying securities that nobody would later buy from them at the prices offered. The focus on process and scale through leverage led them down a path of destroying their balance sheets at a very fast rate. Some very special companies levered up to as much as 1:35, allowing them to borrow $35 for every $1 they had. In a mad rush to beat everybody else, each institution focused on the "volume" and the "size" of their activities. Once again, the focus on process optimization led them to not question if the product made sense. What was the market value of the instruments they were transacting in? How did uncertainty affect value? What was the overall risk in the balance sheet? What may have been the consequences of a sudden and unexpected shock to the economy? Such thoughts were only for mere mortals as the orgy of the great institutions continued till it finally subsided with the subprime meltdown. Those who borrowed money could not repay, the securities that were acquired by the institutions lost value as they were based on the mortgages that went into default, and their balance sheets deteriorated to a point of bankruptcy. Some of these institutions are "kept alive" by subsidies and bailouts and this has a deleterious effect on the overall flexibility of the economy. Failed institutions should be allowed to go away, in spite of their great brands and optimized processes, for the benefit of the economy.

Similarly, contemporary pharmaceutical companies with a focus on efficiency in the R&D process, driven and measured by numerical goals of products crossing certain stage gates, may find themselves developing irrelevant and risky products efficiently. Pharmaceutical companies are perceived to be behind the curve in terms of process optimization compared to consumer goods and other manufacturing companies. To catch up, many pharmaceutical companies are focused on process optimization and this may take these companies further from their goals. Pharmaceutical companies generate information by inventing, manufacturing, and testing chemicals. If this system is optimized for certain numerical outputs such as the number of prototypes, it may result in an efficient increase in numerical throughput. However, since the value added to the company is related to the information content of the prototype, the number of prototypes is an irrelevant metric. A focus on such metrics will reduce the flexibility in R&D and lead the company down a path of lower innovation and increased production of irrelevant R&D prototypes. In such a system, incentives will be designed to meet the throughput goals and not the value goals. These incentives will drive R&D participants to push through R&D concepts and prototypes regardless of their economic value. In this case, the managers of the company may also have an incorrect perception of productivity as the pipelines swell up with prototypes that carry low value or ultimately do not result in any shareholder value increase as they fail at later stages. It may be unwise for companies that depend on innovation as the primary contributor to success to characterize R&D as a process and attempt to minimize costs, maximize speed, or optimize throughput. Instead, they should focus on the characteristics of the products they develop including relevance, economic value, and patient acceptance.

In pharmaceutical R&D manufacturing, companies can increase flexibility by shunning ideas of scale and size. For example, drugs are made for testing in batch processes. Both the active pharmaceutical ingredient (API) and the drug product (tablet, capsule, etc.) are made in large quantities that take significant lead time in manufacturing. The batch processes allow the companies to economize in the supply chain process in terms of ordering raw materials, setting up machines, manufacturing the finished goods, and shipping them to warehouses. A larger batch may allow the cost per unit of the manufactured good to be lower. However, such a process optimization may miss the larger goal. The clinical experiments conducted in animals and humans take smaller quantities of the drugs over a long period of time. These experiments are run over a period of time in stages, each measuring the effectiveness and toxicity of the drug. If R&D manufacturing can produce small quantities of the drug in a continuous manufacturing process, that will introduce flexibility in the R&D process. In a continuous manufacturing process, small quantities of the necessary materials are produced in smaller batches in a process that runs continuously. This may be more expensive, from a manufacturing and cost per unit perspective, than what can be accomplished in a batch process. However, since the lead time to small-scale continuous manufacturing will be low, the overall R&D process can generate information faster. Since materials for experiments are available faster (albeit in smaller quantities), this will help the company to design and run experiments sequentially and generate information that may lead to decisions faster. Since the experiments reveal information, the company holds valuable options to accelerate, modify, or abandon the program. The company's ability to exercise these options optimally depends very much on the arrival of information.

Typically, the clinical and toxicology testing departments drive the R&D process in pharmaceuticals. Thus, process design and optimization always focus on keeping all other functions off the critical path. For R&D manufacturing, this means that they start early and manufacture enough materials so as to make them available to testing functions at all times and at all stages. An early start to the manufacturing process and large batches also mean that materials are always produced in excess of what is ultimately needed. Since the success rates of R&D are low, most programs fail and since materials are made early and in anticipation of successful programs, the result is always large quantities of unused materials. Since pharmaceuticals is a regulated industry and manufactured materials are regulated materials (because of possible toxicity and risk of contamination), they cannot be discarded easily once made. This makes the problem worse as the company, after having made excess materials, will be forced to store and account for them for a number of years into the future.

Now consider a somewhat inefficient R&D process that is driven by the availability of materials in small quantities and continuously. The testing functions can start testing as soon as a small quantity is available. The somewhat inefficient smaller and faster tests will reveal information that will allow R&D to change and possibly redesign the program and in some cases force an early abandonment. As these decisions are made, the manufacturing process can adjust to it. In this case, the R&D does not incur wasted expenses in raw materials that in a large batch process can be significant. If the R&D program is abandoned because of the bad data obtained from experiments, the manufacturing of that specific drug can be immediately stopped and machines can be cleaned and set up for small-scale production for yet other prototype. This flexibility to stop and switch increases the overall capacity in manufacturing and enhances the value of the R&D process. It is not the maximization of production that is important for manufacturing in this case but rather the flexibility to produce just the sufficient amount fast and switch to alternatives as new information arrives. Process optimization based on cost may have been relevant for smokestack companies of the industrial era but they are irrelevant in modern companies coping with high levels of uncertainty and accelerating change. From a traditional viewpoint, both manufacturing and testing are inefficient in this small, fast, and continuous process. However, in this case, the R&D process provides flexibility and it can be shown that the economic value gained is much higher than status quo. Focusing on efficiency for process design can be dangerous in information rich industries such as pharmaceuticals.

Another important process in modern companies is the budgeting process. The budgeting process is typically conducted on an annual basis. Companies budget all aspects of the business such as money, people, time, and space into the various departments or divisions in the company. Most companies follow a rigid budgeting process either based on last year's budget or this year's forecasts. Once resources are allocated to departments and divisions within the company, the managers of each division are held responsible for the budgets. Since the resources required to run the business are driven by uncertainty in the environment, budgets become obsolete as soon as they are put together and signed off. However, the numbers in the budget are still used to measure the performance of the managers. For example, if a division is given certain $ budget, the performance of the division against the budget is largely measured by its ability to "come in on the budget."

Consider a pharmaceutical company where the research and development division has been given a budget, say $X, at the beginning of the year to conduct all activities in R&D. Companies often arrive at this budget using ratios such as a percentage of sales. This year's R&D budget, thus, is linked to last year's sales. However, the resources required to conduct this year's

R&D has nothing to do with last year's sales as pharmaceutical R&D is a long process, often taking decades to bring new products to market. More sophisticated companies may use the current portfolio of R&D projects in setting the budget. However, since the budget is done at the beginning of the year and the portfolio will likely change all through the year (as existing projects are abandoned for technical reasons and new projects added), the actual resources needed will differ significantly from the budget. Once the budget is set, the R&D division may have an incentive to meet the budget, as bonuses may be awarded based on performance against the budget. If the R&D resources required are lower than what is budgeted, because of a thinner portfolio, the divisional managers may then undertake projects that may be lower or of negative value to the company. However, if the spending is on track to exceed the year's budget, managers may slow down valuable projects in the portfolio so as to adhere to an arbitrarily set budget based on information that existed in the past. This deterministic budgeting process and incentives to adhere to irrelevant start-of-the-year numbers introduces rigidity in companies. As the CEO of a large industrial conglomerate remarked once, "budgeting is the bane of corporate America."

To be useful, the budgeting process has to have two important characteristics:

1. It should include the uncertainty in forecasts.
2. It should be dynamic so as to take into account new information as it becomes available.

If a budgeting process cannot be designed to have these two characteristics, organizations may be better off not budgeting at all as providing rigid budgets to divisions in the company create incorrect incentives for divisional managers and they tend to destroy value.

Consider the experience of a hypothetical pharmaceutical company, GiantPharma. In the mid-1990s, GiantPharma was growing rapidly. It had R&D divisions in multiple continents, and it was active in many different therapeutic areas, such as cardiovascular, central nervous system, metabolic diseases, and others. It had a wide variety of functions internally—chemistry, biology, toxicology, clinical, drug substance manufacturing, drug product manufacturing, statistics, and others. Each function and each therapeutic area had a manager and it had a complex matrix system. Rather than providing each manager a rigid budget at the beginning of the year, it created a real time and probabilistic budgeting process. It created a computer program that takes information on all R&D projects in the company, runs a Monte Carlo simulation, and provides a probabilistic budget for each division and therapeutic area. The budget forecasts included probabilistic estimates of people needed, materials to be bought, external expenses to be paid, and space needed for each function. The probabilistic budget was not specific

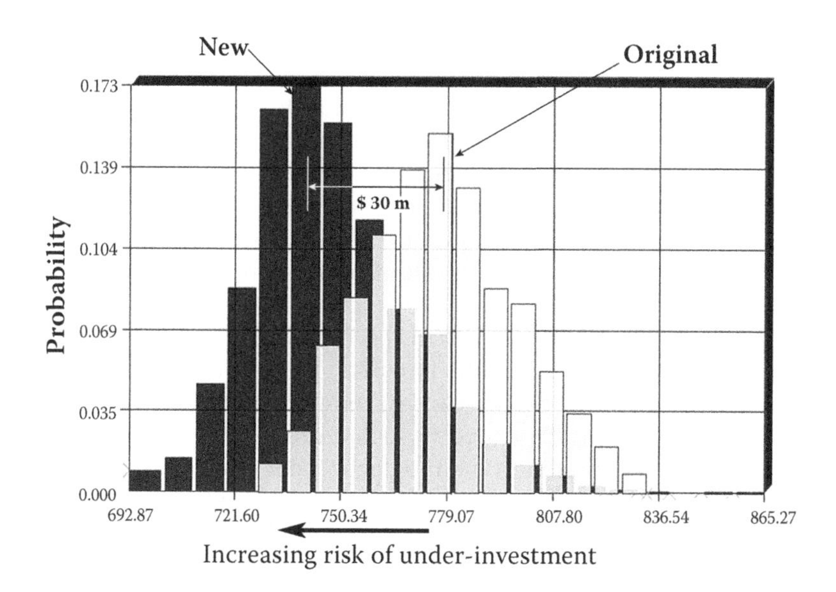

FIGURE 5.1
Probabilistic and dynamic budgeting process.

numbers but rather a probability distribution of possible future outcomes. As new information arrived, the program recalculated the new distributions. Figure 5.1 shows a representation of the output from a probabilistic budgeting process.

The *x* axis shows the resources desired to optimally execute the R&D portfolio in the current year, and the *y* axis shows the probability that the resource estimate is precisely correct. The darker colored histogram represents what the expectations are currently using all the information currently available. The lighter colored histogram shows what the expectations were just last month. As is obvious from the figure, the portfolio changed significantly and the company expects to invest about $30 million less than what was initially planned. In a typical budgeting process, a single number, most likely the average of the lighter colored distribution, will be used in the budget and that will be maintained throughout the year. In environments that exhibit high uncertainty such a fixed and static budgeting process does not provide any information for the managers and will likely create misaligned incentives to spend or not to spend. In a probabilistic and dynamic budgeting process as explained here, the adaptive process becomes an important tool for planning.

Another important process in companies is the earnings forecasting and investor communication process. Certain companies continue to spend large amounts of money and time in precise projections of quarterly earnings per share (EPS), guidance setting, and meeting of the EPS. It is true that if a company precisely forecasts an EPS a few weeks before it is released and fails to meet the forecasted EPS, its stock price will take a tumble. It is not necessarily because the market is "disappointed" at not getting this quarter's EPS, rather it is upset at the managers' inability to forecast and control the company. If the managers of the company "miss" such a tactical measure, the market may lose confidence in them. In information-driven industries that show high levels of uncertainty such as life sciences, energy, and technology, it is nearly impossible to project profits precisely. So, playing the game of forecasting precisely and attempting to meet the forecasts is surely a loser for companies in these industries.

During the recent financial crisis in the United States, many companies suspended earnings guidance and forecasts, primarily because it became impossible to do so. If this change can be made permanent, these companies will save time and effort expended in such a process. A progressive insurance company had done this some time ago. It decided not to provide any earnings guidance to the markets and replaced such guidance with information on the company's business such as new policies and loss ratios. These metrics are much more relevant for the investors and in doing so they have saved valuable time to focus on the real business. There are a few companies that have understood this concept and have decided not to forecast EPS or to give any guidance to the market. They have decided to explain to the market the uncertainties involved in its business and the primary parameters that drive its profitability. This is much more relevant information for the market in projecting the company's value compared to a precise forecast of the EPS. There are a number of value-destroying actions that a company could undertake to meet the tactical EPS—such as shutting down R&D programs, delaying critical purchases, hiring freezes, termination of leases, etc., once the EPS estimate is "promised" to the market. By avoiding such actions and focusing on value-increasing actions, companies can substantially improve their long-term strategic position. It can also be noted that projections of a singular EPS estimate reduce management flexibility and that alone can reduce the value of the company.

There are a number of examples of companies and industries that have lost their dominant positions because of a focus on tactical and static financial measures. Some of this is due to the training managers received from the production economy (where quantity and cost were supreme) applied in a dramatically different world of the information economy that is driven by uncertainty and intangible assets such as intellectual property. Assuming that information companies could be run by tactical financial measures is a gross misunderstanding of how value is created in today's environment. It is the ability of the management to create, nourish, and optimally exercise a portfolio of options that is critical in shareholder value improvement. Good managers get paid handsomely not because they are great forecasters of the future but because they have an innate ability to manage uncertainty. Devising systems, processes, and methodologies to take advantage of uncertainty is the primary differentiator for managers. It is for this ability that they earn their keep.

Content Systems

The characterization of content as a system may be surprising for readers. By *content systems,* I mean the systems deployed by the company to protect and perpetuate its culture, integrity, and viability. The culture of an organization can be codified into a system. It may include how the organization selects, assembles, and integrates employees or participants; how the participants treat, interact with, and communicate with each other; how the organization measures and rewards the participants; and how it interacts with customers, suppliers, partners, regulators, and the environment. In today's companies, much of this is managed by human resource departments with rigid rules. The culture of the company, however, cannot be defined or imposed by a set of rules. It is a belief system shared across the entire organization. It is a framework that allows the participants a method to be consistent at all times—good or bad—and provide them with a sense of belonging. Part of the cultural system is also the language and jargon used by the participants. Companies in the high tech arena have much more defined and pronounced cultural systems compared to other industries. A strong cultural system may help the company shed commonplace ideas in hiring, retention, and firing of employees. Such a company will only attract applications from those who are well aligned with the company's belief system. If employees find that they do not identify with the company's culture, there will also be an implicit but strong incentive for them to leave voluntarily. Such a company can replicate many of the characteristics of the SOUL (self-organizing uni-layer) structure. A cultural system thus is an important enabler in the success of any company.

Companies may have elaborate legal systems that detail what can and cannot be done by its officers, managers, employees, and other stakeholders. Existence of such rules, however, does not mean that the company has a strong legal and moral system. A company that has a set of principles to govern its conduct that is well understood by all participants has a better legal and moral system than the one with defined rules. As the readers may have experienced in the past, much effort and money is expended inside and outside the company to prove and disprove the rules and regulations set forth by the company, regulators, and the government. If the company's moral system is guided by a set of principles, it cannot be misinterpreted by technicalities and exceptions.

Small companies founded by passionate individuals generally show strong moral and ethical systems. As the company grows and is forced to assemble a conventional legal architecture, it loses the clear principles and they are replaced by set of rules that can be interpreted in many different ways. Regardless of its strategy, companies without a strong moral system are unlikely to succeed as the rules will reduce flexibility in the company. The company can adhere to its principles (that are clear) as the environment changes but it cannot continue with a set of rules written in a different environment. Companies should be able to adapt to new environments by applying its principles consistently.

The United States has a high level of specificity in its accounting systems. However, the United States also had many accounting fraud cases in the last several years. Fraud is not a result of lack of rules, nor is it eliminated by a set of prescriptive regulations. Accounting fraud has been more frequent in countries that have more stringent rules and regulations. A set of principles followed consistently will be much more effective in creating a fraud-free environment. As soon as regulators put down a rule, legal experts go to work to interpret it and find ways to circumvent it. Once they find a way to do so, the regulators try to plug the hole by additional rules and the game continues. This does not happen in an environment driven by principles and not rules.

The creation of a strong content system starts with the efforts of the leaders of companies. A company is a reflection of its participants. Weakness seen in the content, such as accounting fraud, starts with the ideas perpetuated by the leaders. The CEO of a company, which perpetrated massive accounting fraud, attempted to defend himself on the pretext that "he did not know accounting." A graduate of a great business school and a high flier in a leading management consulting firm simply forgot accounting when he became the CEO. What cannot be forgotten though is that fraud emanates from the culture of the company and everybody associated with the company is responsible for it. Great companies are not made overnight—they are the result of consistent belief systems and culture and they cannot be created by ignorant and incompetent managers, however smart they are. A true focus on the content systems is a necessary condition for long-term success of any company.

6

Strategy

In this chapter, we will study the third and final aspect of the organization—strategy. Strategy can be divided into internal, external, and boundary subcomponents. Internal strategy is related to how the organization deploys its resources to accomplish its end objectives. External strategy deals with its policies and interactions with competitors, regulators, and the environment. Actions and policies that couple internal and external strategies are called boundary strategies. This may deal with customers, suppliers, partners, and collaborators.

In the 1990s, consulting firms and academics made *strategy* a dominant part of executive management. Creating long-term strategies for the company could only be done by a special breed of people with expertise in seeing long-term trends, speculating on competitor actions and games, analyzing the company's share in segments and its position, creating optimal capital structure, deciding on when and how to act, and many other aspects. Some firms specialized in providing industry analysis and static strategies for companies. This included the rules companies should follow to dominate their industries. Certain firms made strategy formulation fun and easy by equating their clients' products, divisions, and subsidiaries to dogs, cows, and stars and helping them decide which ones to shed, milk, and nourish. Some business schools stopped providing courses in finance, accounting, and other commonplace subjects in an attempt to create world-class strategists by case studies and general knowledge. However, in spite of all the strategies articulated and implemented by companies, most have lost shareholder value in the last 10 years. This could indicate that there may have been some systematic issues with the ideas used in strategy creation. It also points to significant issues with the education doled out by schools specializing in strategy without teaching the fundamentals of business first.

One aspect that needs to change is the static analyses typically done by the strategists regarding prices, costs, quantities, growth, margins, and other aspects and the calculation of a net present value (NPV) by arbitrarily discounting cash flows at some convenient discount rate. Those who shun financial analysis as an untouchable science and attempt to make decisions by superior "gut feel" are not in any better shape either. Four-by-four matrices may have provided sufficient distractions in the 1990s when the economy was growing and many careers were made by the adherence to strategies made from thin air. In the smoke-filled rooms of corporate penthouses, executives of large companies and strategists stood hand in hand in complete awe of their own ability to create strategies—sometimes without analysis and sometimes with analysis supported by massive spreadsheets that meticulously forecasted sales, profits, and margins in every country, segment, and demographic attribute. Many strategists, in love with strategies they helped create, joined their clients, only to realize that strategies concocted in good times may not be the right ones when the going got tough. In spite of the great intellect exhibited by strategists worldwide, in a fast-changing world, it is nearly impossible to internalize all moving parts to reach an optimal gut-based decision. Since decisions are also not static, "gut feelers" have to also consider how and when they may adjust to new information, however smart their guts are.

Deterministic financial analyses cannot also be mended by doing scenarios—a technique practiced by some to introduce uncertainty. However, uncertainty exists in inputs and creating uncertain outputs—such as an optimistic, pessimistic, and average economic impact—does not improve decisions. Those who cannot make optimal decisions on a single number are unlikely do so in the presence of three different numbers. Since long-term strategies do not have a robust feedback loop, those who strategized and helped strategize may have moved on to other jobs by the time the results of their strategies are known. Since strategy does not seem to have much accountability, it has become the favorite hunting ground for those who want to just intellectualize and not be held accountable for the results. The fact that most advisory firms today engage on "fees" rather than for a share of the incremental value created by their efforts is also symptomatic of the problems in the area of strategy formulation and implementation.

Flexibility is typically not an important consideration in strategy development. Thinking about contingent flexibility in decisions was considered only for those who are "weak" and "less confident" in their strategies. In boardrooms and analysts' meetings, it is the confidence of the executive to back a singular and rigid strategy that won admirers. It is the "firm belief" of a leader that he or she would continue forging ahead to the top of the cliff, and even after seeing the canyon below would not stop marching, that differentiated the strong from the weak. It is the ability to communicate in a precise manner how much earnings per share, cost per employee, and margin per product that brought capital providers from all sides, willing and able to fund the company's well-defined strategies. It is the ability to acquire and merge to create larger companies that catapulted companies and their leaders to the top of the *Fortune* list. It is the tactical growth rates (from last year and last quarter) and profits created by cost cutting and operating efficiency that raised many companies to the fortune-teller's list of best stocks to buy. It is the ability to precisely articulate what is to be expected in the income statement coming out the week after, that raised the interest of fast and mad money makers on business television to recommend a buy-buy on their stock. Uncertainty and flexibility are only for the weak, for those who strategize know the future precisely and thus have no use for such artifacts.

Companies also spend time in designing competitive strategies—both speculating what competitors may be doing and will do as well as how to position themselves and their products and services against competitors so as to maximize profitability. Some have ventured into the realm of game theory—envisioning future moves of competitors, suppliers, regulators, and other entities—in a virtual chess game. Game theory has found many admirers both in academics and in business as it renders itself to philosophical discussions, pictures, stories, and general intellectualization in addition to the use of mathematics. It is the theory that appears to be more important than the game, for most games played by companies are based on the needs for today's survival and tomorrow's release of the financial statements.

"Visioning the future" is also common that allows senior decision makers to design future products and services to dominate future markets. The vision of the company leads to its mission and then the mission spawns a set of well-defined strategies assuming a stable and static world. Strategists often have strong opinions as to how the world will be in the future and thus are well equipped to advise companies to design and implement futuristic strategies. Speculation, however, is a dangerous activity in a fast changing and uncertain world. In most cases, forecasting the future is an impossible undertaking, as the world and the players in it make up a nonlinear system with complex interactions leading to unpredictable future outcomes.

Equally prevalent are strategies that "categorize" the company's products and services into boxes such as customer, product, and location segments. Once segmented, strategists can determine optimal pricing, production, promotion, and logistics strategies to maximize the company's profitability. These may be considered "tactical" or "operational" by certain strategists as the last two decades have created a class system in strategy formulators—some willing and able to see "only long term" and some others "experts in strategy implementation." In any case, a rigid segmentation of the company, its products and services, strategies, and other attributes into known industry-wide dimensions may create future constraints. In an interconnected world, common notions of segments may not be suitable. Once formulated, pyramidal structures do not allow easy changes to strategies. In some cases, this leads the organization down a rigid path unable to change as new information becomes available.

Internal Strategies

Let's return to the story of the Aluminum Can Company. As discussed in Chapter 1, this company has a simple business. Buy aluminum and using the established production techniques and facilities to make aluminum cans and sell them. It has two inputs into production—aluminum and energy—and it has a singular output, the aluminum can. The production techniques and machines are in place at a central location. The company ships cans to its customers worldwide. Let's consider typical internal strategies for the company. The operating (internal) strategies of the company may focus on the costs of inputs (aluminum and energy), transformation costs (labor, machines), and the efficiency of production. Some strategists have described the importance of either having a low-cost position or a differentiated product strategy. The thinking has been that the company has to identify its core competencies and focus on that to dominate the market. For example, by creating a low-cost strategy it can compete on cost. As the company wins more business, its scale will increase, further helping it decrease costs of production by taking advantage of the bigger business. On the other hand, a differentiated product will help the company command a higher price and let it move toward premium segments.

Let's consider the internal strategies the Aluminum Can Company could follow to implement the low-cost strategy. To do this, the company has to create an efficient operation and target a high-volume production. It may invest in machines with high throughput and create a factory layout that allows workers to operate the machines and interact with each other efficiently. This will require higher specialization in workers and machines allowing them to operate within limited scope, reducing errors. The company also has to manage inventory of raw materials, components, and products tightly. Supply chains may be optimized to allow low inventory and just-in-time manufacturing and logistics techniques.

In setting up a low-cost operation, a company has to invest into specialized equipment, create efficient and segmented processes to allow worker specialization in activities, create an optimized supply chain with low inventory, and reach high utilization by the implementation of state-of-the-art ideas in just-in-time manufacturing. This all sounds good on the surface, but the primary question remains: What are the objectives of the company? For example, if the company wants to be successful in the packaging industry, it needs to maintain flexibility in its operations to adapt to changes in that industry. Having a low-cost position with high investment and specialization will remove flexibility in its ability to adapt to the industry. It may also reduce the rate of innovation in the company preventing it from influencing the direction of the packaging industry.

The more specialization it has in current equipment, the less flexible it is likely to be. A specialized machine is likely more efficient to produce the specified aluminum cans but may not be configured to do other things. For example, if food manufacturers innovate and create products that require them to breathe air while being inside the package, the company may not have enough flexibility to design and manufacture new products that will satisfy such a requirement.

More importantly, the efficient Aluminum Can Company is unlikely to innovate in the product dimension that will be value-added to its customer.

If consumer taste shifts to plastic or glass, the demand for aluminum cans can drop quickly and the company will not be able to change fast enough. Conventional strategies such as low-cost position have been derived from a financial perspective. Margins and profitability, although tactically important, do not guarantee long-term success and the company's ability to meet its objectives. It is interesting to note that Japanese auto companies in the 1980s had flexibility in many aspects because of their size, culture, and structure. These companies were able to quickly adapt to regimes with very different fuel prices and customer tastes. This flexibility allowed them to continuously improve their market position and eventually dominate the industry. Companies focused on traditional and tactical strategies—such as cost position, customer segmentation, pricing, and other attributes—failed. In an ironic twist, many of the successful auto companies of the 1990s now pursue the strategies of the failed larger companies. There is a high probability that leading auto companies of today will fail in the coming years and will be replaced with lesser-known competitors with higher flexibility to adapt to the uncertain environment.

Many manufacturing companies, aided by strategists, have been shifting production to low-cost countries (LCCs). Conventional wisdom is that cost advantages in these countries will allow the companies to become more efficient and gain a strategic advantage. If such strategies are implemented purely for cost reasons, without considerations to financial and real asset flexibility and their interactions, the company may lose value. For example, a manufacturing company with assembly operations in one country has to consider the interaction of financial (for example, currency) and real (for example, labor shortage, catastrophic accidents, terrorism) risks in its supply chain and incorporate sufficient flexibility to deal with them. If the company finds lower cost of manufacturing for certain components in a LCC, that alone is not sufficient to design a strategy. The additional financial and real risk taken, as well as the possible flexibility gained, has to be analyzed in the context of the entire supply chain. The policies pursued by the LCC are also of importance. For example, if the currency in the LCC is not floating, it removes a certain level of financial flexibility from the supply chain. It may also add the risk of a catastrophic change (such as revaluation of the LCC currency) that increases overall risk of the company.

Standardization is an accepted strategy in many manufacturing companies. For example, automotive companies have been consolidating manufacturing around a few platforms. Implicit in this strategy is that the company expects the public's tastes and preferences of automobiles will continue as in status quo. If the consolidated platforms drive only vehicles of certain size and type, the company may find itself efficient to produce the least wanted products in the future. By defining the company's objectives too narrowly, it runs the risk of complete failure when trends change.

How investors and capital providers view the company can also have substantial impact on the company's strategy and prospects. Small one-product companies are typically supported by venture capital firms, at least partly. The venture capitalist (VC) may have a portfolio of small companies, in effect, replicating the portfolio of a larger firm. However, it is not the diversification benefits that motivate the VC but market inefficiencies such as information advantages related to privately held firms in which they play. However, having assembled a portfolio of privately held firms, the venture capitalist may act similar to the managers of a larger firm (with an internal private portfolio). In these situations, if the venture capitalist does not have a good understanding of the operating characteristics of individual companies, technologies, and products, involvement in operating decisions can destroy value. The trade-off is the information advantage gained by the VC in identifying and investing in private companies against the operating disadvantage of a complex portfolio of not only products but also companies. In this sense, VC firms that focus on the financial aspects of their portfolio and take a hands-off approach to operating aspects may have a higher chance of enhancing shareholder value for their investors. However, the premium that can be commanded by just the intermediation of capital will continue to decline in a well-connected world. Most of the market failures that exist today that prevent small companies from raising capital are related to collusion and clubby behavior. Intermediaries have an incentive to prevent access to owners of capital by "outsiders." In such a system, owners of capital may succumb to the gatekeeper on the belief that it reduces fraud and operating risk for them. Ironically, most fraud happens in well-established, scaled, and well-connected intermediaries.

Let's investigate some of inefficiencies that may exist in small firms due to the behavior of capital providers. Consider a small company with a single R&D program. The budgeting process (using large spreadsheets that take into account all types of costs) may have concluded that the company needs a total of $X million to take the program to completion. This means that the company has to construct an R&D program that has a cost not exceeding $X to prove viability. This hard capital constraint may force the company to run smaller experiments or avoid more costly prototyping. This mode of operation may reduce the probability of success—either because the data gathered is not good enough to prove the hypothesis or the steps skipped (to save resources) results in a failure. So, the artificial capital constraint imposed by the capital provider has increased the probability of failure at the company.

Investing a fixed amount into the company is a lot worse than keeping a flexible and staged budget. Why would someone impose a rigid budget constraint? There are many possible reasons. The capital provider may use the constraint as a "disincentive" on management so they will be more resource conscious. The managers know the implications of running out of money so they will economize. Another reason is downside risk management by capital providers. They may have invested in many similar companies and want to control the risk of overinvestment in any specific company. It may also be "irrational risk aversion"—the investors' risk tolerance decreases after the original investment and the investors start to consider the "sunk costs" in the decision-making processes. They become regretful of the sunk costs and walk away from a valuable asset because they want to limit further losses. Whatever the reason, a hard and arbitrary capital constraint at the time of funding is a sure way to ensure that a company fails.

This type of capital constraint can also exist in large companies, although it is less likely. For example, decision makers may use proxies such as sunk cost as a measure of attractiveness of an R&D program. If the program is more costly than anticipated or slow to progress, decisions to stop may be based on costs, rather than value. Since costs are more "visible" in an R&D process, it unfortunately becomes the only proxy for decisions. The abandoned R&D programs and associated IP also are locked up in most large companies as a sale of these programs to other companies (with differing cost structure, scope, and specialization) is not typically pursued as an alternative for a variety of reasons, including the fear of losing IP from active programs.

Another type of capital constraint is the yearly budget. In this case, the capital provider sets a budget, and if the budget is exceeded, refuses to fund the program (or company) further. This may be another situation in which a capital provider may be using the yearly budget as a disincentive against under-performance or over-expenditure. If the project manager runs out of the yearly budget, she is forced to abandon the project. In this case, the capital provider overfunds the program early and underfunds it later—almost the exact opposite of what may be optimal for the company from a value perspective. A hard yearly budget such as this can also happen in a large company.

Flexibility in operating strategies can be a very important consideration for companies. For example, the operator of a mine has the option to operate or temporarily close a mine, depending on the cost of the mined commodity. This can also be thought of as the flexibility to switch between operating modes—open or closed. Depending on the state of the mine, the switching of the operating modes may incur expenses. For example, opening of a closed mine may incur setup costs, and closing of an operating mine may result in closing costs. Neither opening nor closing costs will be present if the mine was operating continuously. Opening costs may be related to the removal of maintenance equipment, addition of mining equipment and personnel, and other administrative costs. Closure costs may represent preparations to keep the mine idle, employment contract termination costs, and other regulatory and government costs. A closed mine also will involve additional recurring costs, such as routine maintenance because closing does not mean abandonment. In the case of closure, the mine has to be kept in a condition that will allow it to be opened at a future time. There may be regulatory or lease-related costs that may require payments for environmental or local employment reasons. Since these costs exist, the decision to close, to open, or to maintain a mine in any given year is related not only to commodity prices and costs of production but also to the future expectations.

To understand the constraints facing companies in the creation and implementation of flexible internal strategies, consider a pharmaceutical company in detail. The U.S. life sciences industry—a category that includes pharmaceuticals, biotechnology, and medical equipment companies—invests over $100 billion a year on R&D programs that aim to discover and develop new therapies to prevent, diagnose, cure, or alleviate diseases affecting humans and animals. These expenses are spread over a long time, often decades, from the idea generation stage to marketed product. And, with only 1 of over 100 ideas turning into a commercially viable product, the risks are enormous. The life sciences industry (including pharmaceuticals, biotechnology, and diagnostic equipment) generated total revenues of over $800 billion in 2007. The pharmaceuticals segment was the industry group's largest in 2007, generating total revenues of over $600 billion, equivalent to 75% of the industry group's revenue. The performance of the industry group has been forecasted to decelerate, with an anticipated CAGR (cumulative annual growth rate) of 8%. Despite the robust revenue numbers and robust growth forecasts, the pharmaceutical industry has been in a perfect storm for nearly a decade. The factors that have brought it from a business with one of the highest returns on investment (ROI) to one with the lowest are both external and internal.

However, all of these factors have been exacerbated by a management style that is still largely operating as if the environment had not changed. More importantly, with the world's population aging and the risk of bio-terror increasing, life sciences products take on a more important role in the survival of society. To move back to their former glory, life sciences companies have to break away from the shackles of conventional management and traditional ideas. Mergers and long R&D cycles in this industry have led to a lack of accountability and a market test for management decisions. The life sciences industry has to manage many different types of uncertainties—technical, market-based, regulatory, and demographic. The decline in the value of the pharmaceutical sector has now created an acute pressure that needs a response. Clearly, the traditional techniques such as cost cutting through reduction in the size of R&D spending will not fix the problem.

A large pharmaceutical company may have hundreds of R&D programs in different stages. Most of these programs require high investments (i.e., people, money, space, and time) to progress through the R&D pipeline. Since resources are often limited, one of the fundamental problems in R&D is allocation of limited resources to many disparate investment opportunities to maximize the value of the company. This is a complex problem, and different companies approach it differently. Some use the technical risks inherent in the programs as the primary metric for selection and allocation decisions. Others may use minimization of cost, maximization of speed to market, or maximization of overall market potential in making these decisions. None of these ensure that the portfolio that the company selects and advances is the most optimal from a shareholder value standpoint.

Ultimately, resourcing decisions may favor those projects that have strong and vocal champions who can argue for a higher rank for their projects and demonstrate a higher score. The score becomes only a vehicle to show why a project is important. It can also lead to a long and contentious budgeting process as any score can be challenged. An objective criterion such as economic value is less prone to such complications. Also note that traditional NPV is not useful in most cases as it does not capture the inherent uncertainty and decision flexibility in projects. Portfolio management also should be a dynamic process and not something performed at fixed time intervals such as every quarter or year.

The value of life sciences companies is primarily driven by their internal R&D. As previously seen in many other industries, neither acquisitions nor ad hoc cost reductions increase their value. In acquisitions or selective licensing of products from other companies, the price paid generally exceeds value gained and thus can only result in shareholder value loss. Cost reductions, if indiscriminate and ad hoc (as they typically are), result in collateral damage in value-producing entities. In this people and information-intensive industry, cost cutting results in loss of skilled people and lack of overall flexibility for future growth. The market, recognizing this, generally welcomes R&D cost cutting by diminishing the stock price of the company.

The following are reliable indicators that the concept of value is not understood or not applied in a consistent way in valuing R&D and making decisions:

1. Reliance on rules of thumb or proxy-based licensing and contract deals with external partners. This is similar, in many respects, to the valuation practices used in the venture capital industry. Factors such as the size of the overall market, the reputation of management, and any available technical data tend to be the main drivers of transactions. Initial bid prices, which typically take the form of milestones and royalties, are based on those in previous transactions or, more likely, a single successful transaction in the past. There is often some negotiation around such rules, but almost always within limits established by sacred cows (e.g., "absolutely no milestone payment at filing"). Discussions and debates focus on technical details (e.g., "the rat study is very promising, and that is the primary basis of this deal"). How all this translates into shareholder value added is almost never mentioned. Moreover, the split of value between the licensee and licensor created by the deal structure is also not subject to much analysis. The tacit assumption underlying such practices is that if the deal is structured according to the rules of thumb, it is a good deal for the company.
2. Prioritization of programs (and entities) using rankings and multiple evaluation criteria. R&D programs are complex and provide opportunity for multidimensional rankings (on criteria such as

safety, efficacy, manufacturability, differentiation, and the cost of raw materials). Traditional decision trees have been used widely in pharmaceutical companies, and they can be used to calculate the same net present values (NPVs) produced by discounted cash flow (DCF) analysis. But such NPVs are generally viewed by decision makers as only one of several criteria—one that may be too narrowly "financial" to capture the "spirit" of the program under consideration. Moreover, the use of traditional decision trees has led to some confusion in the marketplace as some practitioners have mistakenly labeled it "options analysis" (to denote the branches in the tree). As most first-year business school students know, traditional decision trees are simply pictorial representations of the mechanics of the DCF analysis, nothing more. That is, they do not consider all uncertainties present or incorporate the impact of management flexibility that is inherent in the decision process.

3. Resourcing (budgeting) decisions are segmented by departments, products, and specializations and are often based on last year's budgets. Since there is no common currency to compare investment choices across departments, products, and specializations, resourcing decisions (budgets) are typically done in a segmented fashion. Resources are allocated into buckets, typically according to a formula based on overall sales, last year's budget, and growth rates. Once a departmental (or product) budget is set, managers further divide that amount based on local formulas. Such allocations in turn typically depend on last year's budget or on managers' negotiating skills. It is not unusual to find strong correlations between departmental budgets and the seniority and education of the manager. It is intuitively clear to decision makers that every product or investment opportunity has an intrinsic value to the company. It is also clear that investment opportunities may present various paths forward, and each path (or design) may have different values. If a method could be established to systematically value every investment decision (including alternative designs), one could create a common currency for use in selecting, comparing, prioritizing, designing, buying, and selling investment opportunities. If the method is applicable across all investment choices, the common currency of value can be the only decision variable regardless of the nature, location, time horizon, or size of the investment choices that are available. This is because value, if calculated using an economically consistent method, would capture the information related to all parameters and the uncertainties in the estimation of those parameters. Moreover, the valuation method should be roughly consistent with the intuition and thought process that experienced decision makers go through when they select the best opportunities.

To put this in the right context, consider a pharmaceutical company with the following types of investment choices in R&D:

1. A full development candidate entering phase III (large-scale clinical studies undertaken after proof of concept has been established)
2. An early development candidate that has just filed an IND (Investigational New Drug) application and is ready to enter the clinic (human trials to assess safety)
3. An IT (information technology) infrastructure improvement project that is expected to enhance productivity in record keeping
4. An expansion of a pilot plant that requires high capital expense
5. Hiring of new personnel with specific expertise in oncology
6. A licensing opportunity with a biotech company on a candidate in an area in which the company has its own program.

Suppose also that the company has a hard resource constraint (a limited budget), either imposed by senior management on R&D or set by market forces on the level of R&D that appears optimal at the current time. The question is how such resource constraints should affect investment decisions inside R&D. How does the company decide which investment opportunities to select and prioritize? How and when should the company execute the projects it decides to undertake? In a traditionally managed company, investment opportunities will be selected, prioritized, and funded by different departments in a largely uncoordinated process, and as mentioned, the budgets for those departments are likely to be determined mainly by last year's budgets. Such segmentation introduces the possibility that the best opportunities, if located within the wrong department, may be underfunded or passed over completely for "lack of budget." To make matters worse, finance departments may make tactical adjustments to the departmental budgets to improve quarterly financial statements (apparently believing that investors focus mainly on the next quarter's earnings). As such tactical allocations (cutbacks or increases) flow through departments, they further affect the optimality of the investments undertaken by department managers.

To remedy this situation, companies need to incorporate methodologies and tools that satisfy the following criteria into their decision processes:

1. The methodology is sufficiently generalized to be applicable across the entire investment opportunity set.
2. A tool is available that can be consistently and systematically used across all opportunities.
3. Application of the methodology and the tool is as fast and easy as the application of traditional techniques such as decision trees.

4. Senior decision makers understand both the advantages of the method and the need for change.
5. Application of the tool is sufficiently systematic to be repeatable throughout the organization.

To analyze all investment opportunities in an enterprise, we need a flexible methodology that allows problems to be specified that have both private and market risks. In pharmaceutical R&D, private risks are related to experiments testing the safety and efficacy of the candidates. There may be private risks in R&D manufacturing as well, leading to nonscalability, lack of stability, or inability to manufacture within certain cost thresholds. Such risks should be treated separately and differently from market-related risks. Market risks have to do with the anticipated revenue streams from those products that end up passing the technical hurdles and are actually brought to market.

Boundary Strategies

Boundary strategies deal with the company's buyers, suppliers, partners, and collaborators. For example, companies entering into transactions with others in the buying and selling of intellectual property positions need to consider the uncertainty and embedded flexibility in the valuation and deal structuring. As an example, consider Maxo, a pharmaceutical company that has been active in the area of cancer research for a number of years. It was a strategic shift for the company when it decided to ramp up its oncology (cancer-related) R&D. In doing so, it also invested in sales and marketing of oncology drugs in anticipation of new drugs discovered from its newly focused R&D. As luck would have it, the R&D machine has not kicked into gear yet, and the company is in need of new products to keep its marketing machine busy and its revenue line growing.

One way to do this is to "buy" drug candidates in development from other companies. One such company is Egen, an up-and-coming biotechnology company formed five years ago by a few budding entrepreneurs, scientists, and academics. Egen has been focusing on the discovery of oncology drugs for all types of indications. It has been "firing on all cylinders" and now has three drug candidates in advanced discovery and toxicology testing stages. One such candidate, Edoxin, is getting ready to enter the clinic for the first experiments in humans. Egen has already completed toxicology studies in animal models such as rats and dogs, and everything looks good thus far. On a fine fall afternoon as the Egen managers were conferencing on the development plan on Edoxin, they got a surprise call from the chief scientific officer of Maxo. Maxo has been studying Edoxin after reading an article published by one of Egen's scientists. It looked very interesting to Maxo, and the mechanism of action is something Maxo is pursuing as well. Maxo has a number of discovery programs in this area but none as advanced as Edoxin. Maxo needs a drug soon to accelerate its revenue growth, and Edoxin appears to fill the need.

For Egen senior management, this was a welcome call. As anybody who has gone through a start-up knows, cash is vital. They had venture capital funding when the company started. Further rounds of financing got them to the current stage, but they are quickly running out of money. The company has valuable assets in the pipeline, but unless they can be "monetized," it does not help their cash flow situation. The venture capital firms are always interested in "exit strategies," and some of the early investors are keen on creating internal cash flows to fund newer discovery programs rather than investing more funds.

There may be a motivated buyer in Maxo and motivated seller in Egen. Let us take a closer look at Edoxin. Egen has big hopes for Edoxin and an elaborate development plan. First is the plan to test the drug for a relatively small indication, pancreatic cancer. This is the first indication for which the drug has shown efficacy in animal models. But Egen does not plan to stop there. If Edoxin succeeds in treating pancreatic cancer, the company plans to try the drug for various other indications in the cancer arena, specifically lung, neck, and breast cancer. Smaller companies like Egen economize on development programs by staging them, in this case trying one indication first and expanding it to other indications once the lead indication succeeds and is approved. There are advantages and disadvantages to this strategy. An advantage obviously is committing only small investments early, proving the prototype, and then expanding (substantially reducing the downside risk and increasing decision flexibility by delaying decisions). The decrease in downside risk, however, comes at a cost of reduction in upside potential. One part of the loss of upside potential is due to the delay in getting to market (for additional indications) and the corresponding loss in usable patent life after approval. The composition-of-matter patent on the chemical will be taken during the trials of the pancreatic cancer indication. By delaying the testing and possible marketing of other indications, the company will have less time in the market for those (provided, of course, it succeeds and gains approval). The other disadvantage is that the delay gives the competition a leg up. Once the drug is in the market, competition can study it as well as all the trial data. This may provide them with ammunition to accelerate their own programs, possibly for the additional indications targeted by the company's product.

All these are true, but for a small biotechnology company in a current cash-strapped situation, the best it could do is to push forward with the pancreatic cancer indication and hope for the best. If the drug does get to market, Egen holds valuable expansion options it can exercise or sell. These can be exercised if Egen is successful in raising more money as the purse strings of the venture capitalist may be a little less tight as he or she watches positive cash flows from the approved indication and possibly the widening eyes of the big pharmaceutical companies thinking of acquisitions.

Egen has done an initial analysis of Edoxin's development plan to ascertain the economic value. To do this, it has collected the following types of data:

1. Market size, patients seeking treatment, pricing power, and reimbursement patterns to define the revenue growth. In doing so, Egen has also considered the possibility of competitive entry, regulatory and consumer attitudes toward cancer treatments, as well as the current government policies and saber rattling by certain politicians against pharmaceuticals.
2. Product timelines based on industry norms and benchmarking data. As the company had not done any development programs thus far it did not have any experience-based information here.
3. Cost estimates from a contract manufacturing organization (CMO) to which it has outsourced the manufacturing as it has very limited manufacturing facilities to produce the chemical at scale. They also provided them a manufacturing plan and associated uncertainties such as availability of raw materials, lead times, yield, and other aspects.
4. Design of clinical trials, including the number of patients and dose arms needed based on the hypothesis being tested. In this context, they have also gathered estimates from a contract research organization (CRO), selected to conduct the clinical trials. The CRO has also provided estimates of clinical investigative sites and expected enrollment rates of patients.
5. Based on information gathered from suppliers, the CMO, and the CRO, Egen has modified its project timelines and uncertainties, and also calculated overall costs and uncertainties in various phases.
6. Egen also gathered industry benchmarking data on success rates in various stages for the indications pursued. Their experts then applied some adjustments to the success rates based on the data from the preclinical experiments and discovery chemistry.

Considering all these uncertainties, Egen calculated an economic value for its intellectual property. It has to then consider the milestones and royalty payments it is seeking from Maxo. Maxo analysts calculated a value of the IP position using traditional financial techniques (not considering flexibility and uncertainty). Egen managers had an advantage in negotiations, as they had a market-based estimate of their intellectual property. Using this information, Egen was able to enter a profitable transaction with Maxo, keeping over 75% of the value to itself in a deal that provided generous upfront cash and milestones.

Uncertainty and flexibility are important ingredients in the creation of any strategy. For example, crude oil exploration, development, and production are extremely expensive and involve long processes that carry private and market risks. Although the visible tactical profits of oil companies become the target of public anger and political sound bites, in times of high oil prices, the risks taken and investments needed to explore, develop, and produce from an oil field are often forgotten. Large oil companies that hold leases and other rights on hydrocarbon-rich areas often rely on the expertise and technology of oil service and equipment companies during the exploration and development (E&D) phase. Demands for E&D and production services and equipment are driven by the price of oil and gas. Large companies can provide the entire spectrum of services demanded by the producers, and small firms usually focus on specific aspects of the process.

Equipment such as offshore drilling rigs require long lead times for design and manufacture and have complex supply-and-demand dynamics. As the world hunger for oil remains unabated, over 82 million barrels of "black gold" are consumed around the world every day. With progress in alternative energy stagnating, world growth and stability still depend on the identification, extraction, and conservation of the energy-giving hydrocarbons. Oil exploration is an expensive and high-risk operation. Typically, only governments or large oil companies embark on such endeavors. With oil prices showing high levels of volatility (driven by demand, supply, storage, and perception of remaining stock) and the known deposits of the world's oil shrinking, oil companies find it increasingly difficult to make optimal decisions in E&D.

Let's consider the problem faced by the executives of GiantOil, an oil company that has E&D rights for a large area in western Africa. It won the E&D rights in a bidding round that concluded a couple of years ago and for which it paid handsomely. Although the timing appeared right for the creation of a profitable operation there, GiantOil executives have been hesitant as they felt that the price level of oil (at $100/barrel) was unsustainable. With long lead times involved in E&D, their gut feeling was that initiating the process now will put them at the risk of production commencing when prices normalize (driven by mean reversion in oil prices). Some of the GiantOil executives have been around for a while, and they have had bad experiences in the past by commissioning large projects when prices peaked at significant expense only to find the price of oil, profitability, and the stock price fell when they were ready to produce.

ClearTech is an oil services and equipment company that is considered highly innovative. ClearTech's management has been thinking of ways to enhance shareholder value through more innovative business processes in addition to their ongoing R&D. During a recent executive management team meeting, the chief financial officer (CFO) discussed an idea that found some traction. She suggested ClearTech share the risk of exploration, development, and production with GiantOil. GiantOil has been a long-term customer for

ClearTech, and the management teams at both companies have a long history of collaborating on projects and solving problems together. ClearTech management knew that GiantOil has been hesitant to initiate work in western Africa because of the timing risk and the large investments needed. If ClearTech can share some of the risk, it may induce GiantOil to initiate the investment. Such collaboration also gives ClearTech a direct path to getting involved in E&D in the new area. The project will be large, and it will ensure a steady stream of activity for ClearTech. However, this was new territory for ClearTech since its contracts with customers thus far had been conventional, prescribing fees for services and timelines. Sharing risk will mean that it must settle for lower revenue up front for possible gains in the future. In this case, both the private risks (risk of not finding the anticipated quantity and quality) and market risks (risk of lower oil prices and higher production costs by the time production commences) have to be shared with the customer. ClearTech will have to invest into the E&D process along with GiantOil, and if the expectations do not materialize, their stock prices will take a hit. The CFO was confident that ClearTech could evaluate and price the risk. ClearTech and GiantOil discussed a collaborative agreement to share in the risks. The agreement had the following features:

1. GiantOil will fund some of the E&D costs. It would like to keep these as a fixed amount and let ClearTech pick up the remaining cost. GiantOil felt that the cost of equipment and services are within the control of ClearTech, so such an arrangement will provide incentives for ClearTech to manage E&D costs to a minimum.
2. In the exploration phase, GiantOil will share the cost of exploratory wells but not of the detailed seismic survey. A preliminary seismic survey has already been done, and GiantOil will make the results available for ClearTech. If ClearTech feels that the preliminary seismic survey results are not sufficient, it can embark on a more detailed survey. Since GiantOil will not fund a detailed survey, ClearTech would be responsible for all those costs. ClearTech has multiple ways to execute the project; it can go ahead with exploratory drilling first (with costs shared by GiantOil) or deploy its own detailed seismic survey technology at its own expense to assess the probability of success before embarking on exploratory drilling. ClearTech can abandon the project if the seismic survey results are not good enough to go further. If the exploratory drilling returns dry holes, it can then abandon the entire project as well.
3. In the development phase, GiantOil will share the cost of development and injection wells but not any technology deployed by ClearTech to assess the size and quality of the hydrocarbon deposit. As in the exploratory phase, ClearTech has multiple ways to execute the project; it can go ahead with full-fledged development of the field

(with costs shared by GiantOil), or deploy its technologies to obtain a better assessment of the nature, number, and depth of the development wells to be drilled for optimal production. The new ClearTech technology uses electromagnetic inductance to have a better assessment of the oil field. ClearTech can abandon the project at this stage as well if the assessment indicates that the continuation of the project may be unprofitable.

4. GiantOil profits have to be at some level for it to consider initiating the project. If prices fall below $75/barrel, it will become extremely difficult to manage it. GiantOil wants to share profit with ClearTech as a function of oil price. It will share a larger percentage of the revenue if the price is above $75/barrel. The revenue share will be at a lower bracket if prices drop below $75/barrel.

By considering the various uncertainties in the project and building in flexibility and incentives in various attributes, both companies may have increased economic value for their investors. Both companies focus on what they are good at and keep their options open to make decisions in the future as new information becomes available. The ability to understand various types of uncertainties and design contracts and interactions with collaborators to manage uncertainty better and share risk is an important aspect of strategy.

External Strategies

Competitive strategy gained popularity in the 1990s. Companies strategize in virtually every dimension—businesses, products, locations, and so on. Many of these strategies are static such as the company should enter market X with product Y. To prove such a strategy is profitable, market and financial analysis will show price points, production and logistics mechanics, and competitor actions. To take uncertainty into account, multiple analyses may be conducted—average, optimistic, and pessimistic scenarios—to provide ranges of possible outcomes. The inputs for each of these scenarios are typically fixed. Since multiple scenarios are conducted (with different sets of inputs), the outputs are uncertain. For decision makers, this provides a fallback position. If the company, after studying the analysis, decided to enter market X with product Y and failed, decision makers can point to the pessimistic scenario and argue that they had actually thought about such an outcome. On the other hand, if the company did not go forward but the market proved to be a profitable one for its competition, they can point to the optimistic scenario as the one that actually happened and they had no reason to believe that would have been the case. So, whatever is the decision and whatever is the end outcome, scenario analysis covers all bases for the decision maker, analysts, and consultants. Analysis, thus, is mostly a justification tool for strategists. Decisions are generally taken on gut feel and then many analyses are commissioned to justify them.

However, creating uncertain outputs using constant inputs is not a useful exercise since this does not provide any decision guidance. In creating strategies, the company has to own up to the fact that all inputs (market size, price points, costs, timelines, competition, regulation) are uncertain and hence methodologies that evaluate uncertainty are needed in crafting effective strategies. The ability of the company to enter market X with product Y is an option the company holds. Such an option may have an optimal exercise horizon. In many cases, waiting is not necessarily a bad strategy. Any strategy that increases (or maintains) flexibility may be valuable for the company. Once the company enters market X with product Y, it loses some level of flexibility. That is not to say that the company has to always wait but it has to consider the loss of flexibility associated with the execution of the strategy into account.

As an example, consider a fashion garment producer. Styles and consumer preferences change routinely. It is difficult to precisely predict what fashions, styles, fabric types, and color are most popular every year. If the company specifies all attributes of a new clothing line it plans to introduce next year, it can produce them more cheaply, as this will allow the company to provide longer lead time and commitments on higher volume to suppliers of components. However, if the attributes of the clothing line are fixed early, the company loses all flexibility and runs the risk of producing less popular products. If the company delays the decision, it maintains the flexibility but its costs will be higher. Another alternative is to produce various types and colors in small quantities and test them. This staged approach will provide the company with additional information as to which designs are likely to succeed.

Strategy is also about finding an optimal configuration for the business the company is in. For example, consider a technology company that has been considering spinning off one of its divisions to raise cash for future investments. This division, which manufactures mature electronic components, has operations around the world and an infrastructure that is well managed. The decision to spin off has been contentious among senior managers. Some felt that the division is well positioned for a spin-off. They made the following compelling arguments:

1. The new company can provide good transparency into its source and use of cash for investors and thus allow it to raise financing if needed in the future.
2. The business is reasonably mature and as a separate company it may be able to increase innovation and create new products to drive growth in the future.

3. The spin-off allows the parent company to raise cash quickly and redeploy it to other areas such as expansion of its core products to emerging markets.

However, a smaller contingent felt that there are synergies between the components division and the rest of the company, including controls and robotics. They argued that many of the design engineers work across the divisional boundaries and share ideas. Further, a detailed understanding of component costs allows engineers to design more efficient controls and robots, the company's core products. After many months of debate, the management team was nowhere close to a final decision, and the investment bankers they engaged to advise were getting impatient.

One of the analysts in the corporate finance department had an interesting idea to possibly bring the two sides together. She suggested that the company consider a "carve out" of the division with an option to buy back control if the "synergies" that were argued to exist were significant. In effect, the company can spin off the division but hold an option to buy back 51% of the equity (to gain control) at a predetermined price within a period of time. This is a simple call option. To effect such a transaction, the company may have to pay the investors something close to the value of such an option. It may be able to do this by providing a discount to the initial price. In a typical carve-out, only a minority stake in the new company is provided for the IPO, and the parent company holds the majority stake. In this "special carve-out" suggested by the analyst, all the equity is provided for the IPO, but the parent company holds an option to buy back a majority stake in the future. This option is valuable and may have an effect of capping the returns to the IPO investor.

The mechanics of the option exercise will be as follows. If the parent company exercises the buyback, the new company will issue new shares that will be bought by the parent company at a prespecified price. The parent company is likely to exercise this option only if the price of the stock of the new company is higher than the prespecified strike price, and this acts as a cap on the stock price of the new company. Hence, the exercise of this option may mean dilution to the existing shareholders of the new company. However, if the option is priced properly and a discount is given in the carve-out equity, this can be a win-win for the company as well as the new investors. Ability to consider uncertainty systematically and utilize flexibility to create better alignment of incentives can add value to companies.

Managing uncertainty is a big challenge for multinational companies. Companies with operations in many different countries face exchange rate exposures. There are multiple types of exposures, such as translational, transactional, and operating exposures. Translational exposure results from translating revenue and costs that occurred in another currency to the home country (or reporting) currency. It is a tactical accounting exposure. Transaction exposure comes from contractual terms expressed in one currency and the movements in the exchange rates by the time the contract is executed. This is generally short-term in nature. Operating exposure emanates from the organization and operations of a multinational company that generates revenue and incurs costs in many countries as the result of the company's strategy. Operating exposure is a long-term issue for any multinational company. Transaction and operating exposures are related and may have to be considered together.

Financial hedges are typically used to ensure against short-term moves in exchange rates. An airline's purchase of forward contracts on fuel is an example of managing operating exposure, but such contracts can also be short-term, ensuring against anticipated transaction exposures. Since such contracts have a fixed timeline, changes in underlying business (as in the case of a rapidly falling demand for air travel) may expose the company to exchange rate risk in a direction opposite what was carried prior to entering a forward contract. The longer the timeline associated with the exposure, the more difficult it is to manage such exposure with more tactical financial hedges. Even the best thought out financial hedges can ultimately result in exposures in a direction not anticipated by the company. Such events can be extreme and may even result in the failure of the company.

To effectively manage currency exposures, multinational companies have to design and manage an estate of decision options (strategic and operating strategies) in a fashion that maximizes the value of the firm. An example of a decision option may be Toyota maintaining flexible manufacturing operations in Japan and the United States, allowing it to switch production to take advantage of exchange rates in the future. It is not only the more obvious location flexibility that is of importance here, but also other important aspects such as equipment flexibility (ability to switch production lines to different automobiles such as cars and trucks), labor flexibility (ability to stop and start production by ramping employees down and up), equipment/labor switching flexibility (ability to change labor content in the manufacturing process by incorporating more or less automation), energy flexibility (ability to use a variety of energy sources), and others. Flexibility may also be designed in other parts of the operations, such as engineering, marketing, and logistics. The engineering design of manufactured goods should be a focus area for companies as the design can enhance the company's ability to take advantage of various types of flexibilities that may exist in the supply chain. For example, modular design and switchable components could provide the company opportunities to disaggregate manufacturing on an as-needed basis and the ability to shift production to different locations. Postponing final customization is an example of enhancing logistics flexibility. In this case, the company delays the final customization of the end product until it is ready to sell the product wherever demand is available. The uncustomized product can be shifted from a low-demand area to a high-demand area, thus enhancing marketing flexibility as well.

Country-based branding coupled with regional brands that target a wide range of price points is an example of marketing flexibility that allows the company to move up or down the price/brand spectrum on an as-needed basis. If an automotive company proactively takes actions to establish a production network with such flexibilities designed in, its ability to take advantage of tactical exchange rate fluctuations increases, and the need for pure financial hedges (which can be expensive) decreases. Because exchange rates are mean reverting, there is a "convenience yield" to actually having a flexible operating strategy that is active rather than "on paper." The same is true for other types of flexibilities, such as in engineering, marketing, and logistics. The lead time in implementing flexibility may be reduced by a financial transaction such as buying a company or existing assets of another company. However, in such a case, the company may have to forfeit (or the very least share) the value generated by flexibility as the seller is fully aware of such value.

Collaborations and partial equity stakes in other companies in locations where the firm anticipates exchange rate exposures are other ways to increase flexibility without high up-front investment. One could argue that such actions also provide the company an option to delay larger investments. However, this may also have a price as the collaborator may seek features in the contract that may deter the firm from taking advantage of favorable future exchange rate regimes by abandoning the collaboration and going on its own. It is often the case that corporate finance focuses on the financial aspects of the company, including the tactical financial hedges, and other departments such as strategic planning worry about decision options. Such management configurations will certainly suboptimize the value of the firm, as each will attempt to maximize only one piece of the puzzle. To make matters more complicated, there can be many interactions between financial hedges and decision options in the company, and systematic portfolio management of the entire estate of financial and decision options are the only way to maximize value.

Traditionally, economists have argued that a firm should "hedge" all risks over which it does not have control. This will allow the managers to focus on those aspects of the business that they can influence. However, such a thought process may make managers focus only on tactical financial hedging. Firms may not only have to reorganize (away from traditional silo management of financial and real risks) but also implement tools capable of analyzing all risks in the same framework. Such tools should also be able to incorporate interactions between risks. In this case, firms may also identify risk components that reinforce each other and when certain thresholds are breached may exhibit characteristics that cannot be controlled. It is this "runaway train" phenomenon and not the lack of traditional risk management that subjects most companies to the risk of failure. Tactical and financial risk management may keep them afloat when risks are not high, but such a process may not be enough when it is subjected to a sudden shock.

Clearly, the scope of the firm is important in its ability to take advantage of a systematic management of financial and decision options. The larger and the more spread out the firm is, the higher the value of such a process. The currency regime in which the company is operating is also a factor. If the home country's currency is pegged and the firm largely transacts in its home country, it may not have a great reason to think about exchange rate exposures. If the firm's products and services have more commodity-like characteristics, it is more likely that it can take advantage of flexible manufacturing and logistics operations as their location specificity will be low. Even those firms with specialized products can move down from specialization to commoditization by breaking their products down to components and increasing configurability and switchability. Such a thought process, seeking flexibility in every aspect of the firm, has now become a necessary condition for survival and success in a connected world of people, products, and skills.

For many firms, a substantial part of strategy has to do with dealing with governments and other nonmarket entities. Unfortunately, even the free market bastion of the world, the United States, is moving in the direction of government ownerships of incompetent companies. As an example, consider a construction company bidding on an infrastructure project. Some developing countries, such as India and Brazil, are in need of infrastructure development and improvement. The government can entice commercial enterprises to partake in such activities through a combination of subsidies, guarantees, and other incentives. These contracts allow flexibility but they also come with uncertainty. One such project is a build, operate, and transfer (BOT) scheme for highway construction connecting two cities. The government awards a BOT contract to a commercial enterprise, typically through a bidding process. Let us analyze such a project. The project may have characteristics such as the following:

1. At the end of construction (build phase), the company has the option to sell the highway back to the government for a fixed price of $X million.
2. During the build phase, if the company abandons the project, no repayment is possible, and the government will take over the project.
3. If the build phase takes more than certain number of years, the company has to pay the government a penalty of $Y million for every additional year.
4. During the operating phase, the company can collect tolls for the traffic on the highway.
5. At the beginning of each year of the operating phase, the company has the right to demand its debt service cost in lieu of tolls for that year.
6. Every year during the operating phase, the government will guarantee tolls of certain value. If the tolls collected is less than this guarantee the government will pay the difference.

All of these are options in the contract. By systematically considering all uncertainties, the company can value each of these contract features and optimally bid for the contract. By understanding the value of each of the options and creating a "reserve price" for the contract allows the company to avoid "winner's regret." In projects such as these, bidders without a clear understanding of the economic value including the suggested options may overbid, win, and ultimately lose money.

In summary, consideration of uncertainty and flexibility is essential in crafting external strategies. Many of the traditional strategy formulation processes use unchanging data. Scenario analyses may be conducted to provide bounds to outcomes but having such bounds does not help in picking the right strategy for the company. If decisions have contingent future flexibility, the company will make wrong decisions by assuming that all future decisions need to be made when it makes the current decision. It will be unable to consider the various options to delay, accelerate, abandon, and switch as new information becomes available in the future. Strategy formulation and implementation is fundamentally the creation and optimal exercise of options.

7

Flexibility Metrics

In this chapter, I will construct a diagnosis kit to identify flexibility-related problems in companies. A company's current stock price is a reflection of the market's expectation of the future prospects of the company. The characteristics of historical stock prices may give us some indication as to what has been happening in the company. What we are interested in is not predictions of stock prices but rather the "shape" of the stock price returns from the past. For example, we can measure the uncertainty seen in the stock prices of the company in the past. We can measure the uncertainty in stock price by its *volatility*. This indicates how much the stock price moved around in the past (historical volatility). Companies in volatile industries such as high technology, life sciences, and energy are expected to show high volatility in their stock prices. This is because these industries are changing fast and new technologies and innovations appear routinely that have a big impact on the companies that play in these industries. We can calculate volatility by using the daily, weekly, or monthly returns from the company's stock price and finding a standard deviation of those returns. The higher the standard deviation, the higher the volatility and the overall risk exhibited by the stock price. If the company's stock has options trading in the market, another measure, called *implied volatility*, can also be calculated. This is possible because volatility is an important factor that drives the prices of options. The implied volatility shows how much the market expects the stock price to move around in the future. So, this is a forward-looking measure.

The volatility in a company's stock price, however, can emanate from two different factors. First, it may be due to the overall volatility of the industry, country, or economy, and second, it may be due to the unique characteristics exhibited by the company. Finance professionals call the former *systematic* (market) risk and the latter *unique* (private) risk. The overall volatility exhibited by a company's stock price, thus, is a combination of market risk and private risk. The private risks are specific to the company and include risks such as patent infringement and the failure of an R&D program. The market risks emanate from the market and indicate how much the company is exposed to the overall economy. For example, a company with high market risk will do poorly in recessions and really well when the economy is growing.

Readers may be familiar with a financial metric called *beta*. Beta helps us measure the market risk in a company's stock price. Beta is a measure of how correlated a company's stock is and how volatile it is compared to the overall market. So, beta depends on two factors—the tendency for the stock price to move along with the market (correlation) and the variability in the stock price returns (volatility) in comparison to the variability in the overall market. If a company's stock price return shows high correlation with that of the market, it means that the company is less able to mitigate the shocks emanating from the broader economy. These shocks may be unrelated to the company and its business (such as changes in tax regimes, wars, etc.). However, if the company's stock price is moving in lockstep with the overall economy, the company has less flexibility to mitigate macroeconomic shocks. Correlation of a company's stock price returns to the broad economy can thus be an indication of inflexibility. A flexible company's ability to dampen the shocks emanating from the economy will result in its stock returns showing a lower correlation with that of the market, and this can result in a low beta. However, flexible and innovative companies can still show high volatility in returns as they bring new and great ideas to fruition and face periods of failures. These company-specific events could occur any time and will be uncorrelated with the economy.

Thus, if a company's stock price is rising as the market is rising and vice versa, the company may not have much flexibility. The managers of these types of companies may be paying themselves huge bonuses as its stock price rises during a general rise in economic activity even though they had nothing to do with it. Ironically, in many companies, when the company's stock price falls along with an economic recession, managers are not paid negative bonuses (or asked to return their bonuses from earlier years). Instead in many companies, bonuses continue albeit at a reduced rate and the managers will then blame the downturn in the macroeconomy for all their troubles. In general, traditional managers who take credit for themselves as stock prices go up in an expanding economy and blame the economy when stock prices go down in a recession have no ability or desire to increase flexibility in their organizations. If a company removed all flexibility in its structure, system, and strategy, it will be exposed to the macro-uncertainty directly with no ability to manage it. Thus, we can use beta and volatility to categorize companies. Uncorrelated volatility (random volatility) is better than correlated volatility (stock moving in tandem with the market).

However, not all uncorrelated volatility is good. If the volatility (standard deviation of returns) is normally distributed, it may indicate that the company is on autopilot—they succeed sometimes and fail in other times and the frequency of successes and failure are approximately equal. It is a bit like flipping a coin—you win sometimes and lose sometimes. In this case, the company may be conducting its business in an ad-hoc fashion—sometimes doing well and sometimes doing poorly. To take this one step further, we can look at how skewed the company's returns might be. Skewness is a measure of departure from a normal distribution—a positive value indicating a longer positive tail as shown in Figure 7.1.

If a company has flexibility, it is more able to create a positive skew to its returns. This is because flexibility allows the company to limit the downside

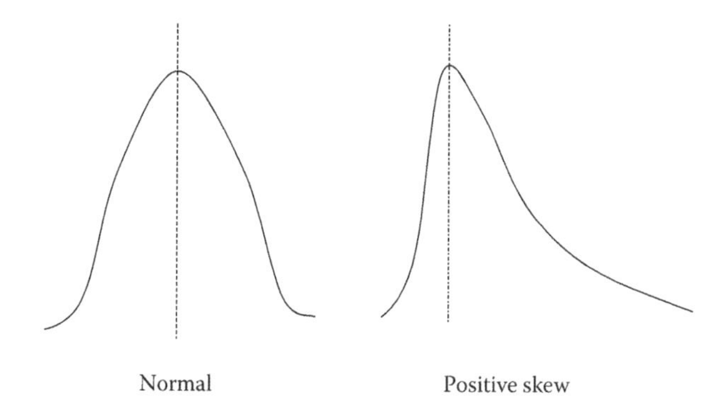

FIGURE 7.1
Normal and positively skewed probability distributions.

risk while enhancing the upside potential. For example, the Aluminum Can Company with infrastructure flexibility in machines can configure its plant to take advantage of high demand for certain type of aluminum cans, increasing returns in good times. If the demand drops, it can switch into other types of products, limiting the downside losses. Flexibility, thus, allows companies to exhibit a positive skew in returns to investors.

Positive skewness, however, can also happen by random luck. For example, in the 1990s, many venture capital companies showed high positive skew in their return distributions. Many believed that they were extremely good as they have been able to identify the real high fliers. Over time, it became clear that a large percentage of the successes in the 1990s for many companies came from pure luck rather than skill. This is also true with money managers and hedge funds in the public markets. Typically, the list of best-performing mutual funds, money managers, and hedge funds all change from year to year. Those who do well this year are not the ones in the best list next year. This lack of consistency implies that it is more driven by luck than skill. When managers do well they advertise and pick up clients and when they do not, they may simply keep quiet and wait for the next year. Mistaking luck for skill is a common problem in many industries. Negative skewness in the returns means that the company had extremely bad events, albeit with low probability. It also implies that it had more bad returns on average than good returns. The managers of these types of companies may be destroying value by making changes when none may be needed. Their presence in these companies actually reduces value for the shareholders. Removal of managers and anybody who is not needed for the operations of the company and introducing more automation may help these companies move back to a normal shape of returns.

Figure 7.2 shows a schema for the identification of companies using the shape and characteristics of their stock returns.

High positive skewness	Flexible company	Unknown
Low or negative skewness	Machine	Rigid company
	Low beta	High beta

FIGURE 7.2
Categorization of companies in the skewness-beta matrix.

Assuming that the skewness of returns in a company's stock is resulting from the application of flexibility, we can use this to categorize companies. If high beta is coupled with low or negative skewness, it is the worst combination. These may be rigid companies that move and up and down with the economy (high beta) and lose additional value due to bad managers (low or negative skewness). We will call these types of companies "rigid." The managers of these types of companies destroy value in two ways. By setting up a company that simply mirrors the economy, they have created a rigid company. Additionally, by meddling with the company's affairs, they created a negative skewness in returns—perhaps taking excess risks at inappropriate times or not optimally exercising options when they were available.

If the company shows low beta and low skewness, it may mean that the company is in a counter-cyclical business that naturally shows a low beta. It may also be due to fixed contracts and infrastructure that effectively fix the company's profitability regardless of the state of the economy. However, managers may attempt to "manage" them and in the process lose value as shown in the negative or low skewness. We will call these types of companies "machine," to indicate low volatility and low management competency.

On the other hand, if a company is able to demonstrate low beta and high positive skewness, it may have flexibility. Low beta and positive skewness show that the company is able to manage uncertainty well through flexibility. The company has been able to take advantage of uncertainty by creating valuable products and services with high upside potential and in bad times it has been able to limit the downside risk. The successes and failures of its products and services happened uncorrelated with the market, indicating it is due to the actions of the managers. We will call these types of companies "flexible."

If a company shows high beta and high skewness, we will characterize them as "unknown," since it is difficult to determine the dominating characteristic, as they have opposite effects.

Figure 7.3 is an example of a set of U.S. companies in a skewness-beta matrix.

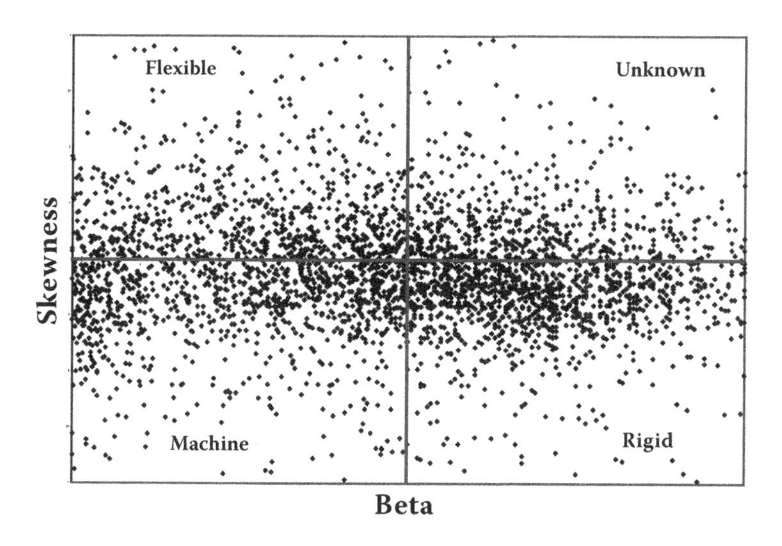

FIGURE 7.3
A sampling of U.S. companies in the skewness-beta matrix.

Now let's investigate overall metrics for the value generated by a company. The value of a company or organization emanates from many different components including its marketed products and services; its intellectual property and ongoing research and development that may result in new products and services in the future or enhancements to existing ones; its structure, systems, and strategies that provide uniqueness and flexibility to the company; and many others. Some of these are tangible and visible. For example, a consumer goods company may have marketed products under well-established brands. These products may have been in the market for a while and so the cash flows associated with them are easier to predict and evaluate. The expected growth in these cash flows can also be assessed based on the growth in the current markets as well as new markets the company may enter. Similarly, a telecommunications services company can reasonably predict the revenues associated with existing products (say, land-based telephones) based on the existing customer base and predicted growth. These cash flow projections may not be precise but both the consumer goods and telecommunications services companies will have a narrow band around their projections because the products and services are well established and are in known markets. The value emanating from the established products and services may be calculated using traditional financial analysis such as the discounted cash flow (DCF) analysis. Typically, the company will project the cash flows and then discount them back to the present to account for time value of money and associated product risk.

On the other hand, if the consumer product company is engaged in research and development for new products, the cash flows associated with the possible future products are more difficult to predict. There are a number of different uncertainties around research and development—such as unknown costs, timelines, technical risks, and eventual market potential—that needs to be considered in valuing such intangible assets in the company. Also important is any decision flexibility that may exist in conducting the R&D. For example, R&D programs may be staged (i.e., conducted in phases), and that allows the company to learn about the technical feasibility, costs, and market acceptance as it progresses in time. Such a staged R&D program may offer valuable options to the company for the delay, acceleration, abandonment, and modification of the program as new information arrives in the future. This increased flexibility coupled with the uncertainty in the various aspects of the R&D program makes it difficult to be assessed with conventional tools and techniques. Options-based analysis techniques can be used to value them, however.

Similarly a telecommunications services company will have uncertainty and decision flexibility if it is designing a new wireless offering. It may be collaborating with mobile phone manufacturers in designing the program and rolling it out. There may be uncertainty around customer acceptance of the new service. The company may create newer service and product offerings as it understands competitive positions and customer tastes and creates pricing schemes that may bundle or unbundle services and products to better serve specific customer segments. This uncertainty coupled with the flexibility to modify the program later provides the company with valuable options. There may be other intangible items in a company. For example, a company with a strong brand and culture may find it easier to recruit, retain, and motivate employees. A company with a flat and less bureaucratic organizational structure will be more flexible and innovative, enhancing value.

The value of the organization, thus, can be divided into two major components—one that represents tangible value and the other that represents intangible value. The intangible value is driven by the flexibility the organization has to take advantage of the uncertainty it faces. Both uncertainty and flexibility have to exist for intangible value to be present. If there is uncertainty but the company has no flexibility it cannot take advantage of the uncertainty. If flexibility is present but the company is not facing much uncertainty, there is nothing to take advantage off. Hence intangible value happens in the presence of uncertainty and flexibility together. From a financial perspective, we can represent the total value of the organization in the following fashion:

$$\text{Value} = \text{DCF} + \text{FLEX}$$

where

DCF = discounted cash flow value (value from established products in known markets or from those with no flexibility or uncertainty)

FLEX = Value emanating from future options (intangible value) or current products that offer uncertainty and flexibility.

Most of the value of the traditional companies resides in the DCF component. The DCF value can be calculated by projecting the cash flows of the company's known products and services and discounting them to the present. This idea comes to us from the capital asset pricing model (CAPM) that was put forward in the 1960s. Most of the top executives in today's companies grew up with the CAPM and have come to have great faith in the DCF approach. Sometimes, instead of a discounted cash flow (DCF)–based net present value (NPV), managers may calculate an internal rate of return (IRR) and compare that against the cost of capital. Both of these approaches are based on the financial theory of CAPM. The implementation of this, however, differs from what the theory teaches.

For example, the theory teaches that the discount rate used in the DCF analysis should reflect the systematic (market) risk of the project/product. The value of a company is the sum of the values of all of its products and services. Since companies have different products and services and each may have a different market risk, it has to value each product and service using a different discount rate, reflecting appropriate market risk. For example, for a telecommunications services company, the discount rate for the wired telephony product may be lower than the discount rate for optional services in the mobile market. This is because households may consider the landline to be an essential product and the optional services in the mobile space to be truly optional. In this case, consumers will continue with the landline even if the economy enters a recession but may chose to terminate the optional mobile services. Thus, the optional mobile services will be much more closely correlated with the state of the economy and will command a higher discount rate to reflect the higher market risk. In effect, the discount rate should be based on the beta of each of the product and services the company offers. Since the company is a sum of all of its products and services, some more risky and some less risky, the discount rate for the company will be the average of all of them adjusted for any financial leverage. Also, the cash flows have to be adjusted for technical risks as these risks are not market driven and the discount rate should only include market risk. The calculation of an internal rate of return (IRR) based on projected cash flows and comparing that against the average cost of capital thus misses the underpinnings of the capital asset pricing model. To make matters worse, the younger generation may consider the landlines to be a luxury item and mobile services to be essential. So, the discount rates for products may also change with time.

Most fundamental analysts who project the value of companies and thus their stock prices use similar approaches as the DCF (both correct and incorrect applications of the traditional theory). However, the value of a modern company is mostly driven by the value of flexibility (FLEX). These are related to future innovations in the company or the decision flexibility the company has to alter future cash flows that are uncertain. As is evident from the recent financial crisis, the valuation approaches followed by the fundamental analysts and financial institutions, largely driven by DCF, was flawed. These companies and the world economy are paying a high price for valuations based on the blind discounting of the projected cash flows. The overall market, however, is able to understand the FLEX value of companies and the stock prices of companies typically reflect this expectation. Financial analysts wrap this into ratios such as price to earnings (P/E) in an attempt to reflect the market's expectation when their own DCF models do not provide reasonable guidance.

To understand the two components of value—DCF and FLEX—let's revisit the Aluminum Can Company. Let's assume that the company currently has a single product (soup-size can) and a single customer (soup manufacturer). To calculate the DCF component of value, financial analysts will typically forecast the volume based on industry trends in soup consumption, market size, market share of the end customer, and other factors. They may also consider the cost and pricing structure of competitive producers of soup cans and the existing contract between the Aluminum Can Company and its singular customer. Based on the industry and competitive dynamics, they may establish a pricing power for the company and deduce a price per unit. On the cost side, they may analyze labor and energy costs, cost trends, inflation, and other attributes. Such an analysis eventually leads to a constant set of cash flow projections into the future. For example, they might establish that two years from now, the company will sell q cans/year at a price of \$$p$/can and it will cost them \$$c$/can to manufacture and ship. Thus the pretax cash flow for the company will be $q \times (p - c)$. If q, p, and c are determined precisely, this cash flow will be a singular number. This can be done for a number of years into the future (they may establish a life of the company or determine an end value based on the length of the contract). Once the cash flows are determined, they can be discounted back using a rate that represents the risk of capital deployed in the aluminum can business. The DCF value thus captures what is known to occur—production and sale of soup cans to its long established customer, the soup manufacturer. Both the internal and external financial analysts conduct processes similar to this one and calculate a value for the company and its stock price (if the company is publicly traded).

In most cases, the DCF analysis conducted by the analysts will not agree with the observed price of the company's stock in the market. This often prompts two types of revisions to the analysis: (1) an optimistic or pessimistic forecast of cash flows is done and corresponding DCF value is calculated so as to bracket the observed stock price, and (2) a comparative analysis is done by comparing the company's P/E (price to earnings), P/S (price to sales), or P/B (price to book) ratios to its competitors or industry averages. These types of analyses may still show a discrepancy that may prompt the external analysts to recommend a buy or sell on the stock or the internal analysts to communicate the true value of the company to the market through analysts meetings and other such vehicles.

In the case of the Aluminum Can Company, let's assume that the stock price of the company is roughly equal to its calculated DCF value. What does this imply? This means that the value of the company is driven primarily by its current product and customer. The investors in this company expect a series of cash flows based on the sale of its soup cans to its only customer. In other words, the investors do not believe the company will have any new products or services and it acts like a machine that is constructed to do one thing. In the modern world, such companies are akin to pure machines as technology allows us to automate most of the routine activities today. It does not require many employees nor does it require any managers. The machine does not have any uncertainty to deal with (as volume, price, and cost are forecasted to be static) and no decisions to make. However, in practice, the company has to deal with uncertainty in cost, price, and volume (even if it has only a single product and a single customer). If it is unable to manage this uncertainty because of inflexibility in the structure, system, and strategy, the managers of this company can only destroy value and not increase it. If this is the case, the market value of the company will be less than what can be expected from the DCF analysis. The difference is the value destroyed by the leaders of the company through inflexibility and bad decisions.

If we find that the market value of the Aluminum Can Company is greater than the DCF value, two conditions have to be true.

1. The company faces uncertainty. Future cash flows are uncertain.
2. The company has some flexibility to alter the future cash flows—through better management and innovation.

Thus, the difference between market value and DCF value is driven by two components—uncertainty and flexibility. Both of these have to be present for the FLEX component to have a positive value. If uncertainty is present and the company has no flexibility, the FLEX component can also be negative. If the FLEX component is negative, the best action for the company's managers is to close down the company and return everything back to shareholders as it implies that managers are destroying value. Higher uncertainty increases value (if flexibility is present) because it provides the company more opportunities to increase the upside.

Conceptually, we can categorize companies in the schema shown in Figure 7.4.

If a company has low DCF value but high FLEX value, it is likely an emerging company with flexibility and innovation capability. The company does not have many products in the market nor does it have high revenues and profits. The market, however, indicates that the company has great potential and its inherent flexibility is expected to create great products and services

	Low DCF value	High DCF value
High FLEX value	Emerging high flexibility	Balanced portfolio mature but flexible
Low FLEX value	Dying	Mature and rigid low flexibility

FIGURE 7.4
The FLEX-DCF value matrix for categorization of companies.

in the future. Most companies start in this fashion, aided by great ideas and a group of people who work together well.

If the company has both high DCF value and high FLEX value, that indicates the presence of a balanced portfolio. The company may have been around for a while as it has a robust stream of profits from established products in the market. In addition, it seems to have been able to maintain sufficient flexibility as demonstrated in the FLEX value. The market thinks that the company will continue to create new products and services in the future to replace or complement existing products and services. It seemed to have avoided some of the pitfalls of growth and the market is still hopeful of continued success at the company.

If a company shows high DCF value but low FLEX value, we can surmise that it is a mature company with many established products. It, however, has little flexibility as the market does not expect the company to innovate and create future streams of value. Most large and old companies are in this category. They are riding a wave of product introductions that happened in the past, when they may have been flexible and innovative. Buoyed by the high profits created by these products, these companies may have grown fast, hired indiscriminately, bought other companies, created inflexible and large buildings and plants, created complex divisions and subsidiaries that are not integrated, and followed strategies based on rosy and unchanging future forecasts. This story is played over and over again in the history of companies that were founded by visionary entrepreneurs, innovated, and then succumbed to the templates of scale, efficiency, and inflexibility.

If a company shows both low DCF value and low FLEX value, it may be in a dying phase. It may have been a young company who could never make it big or it may be an old and mature company that ran out of flexibility and ideas. In either case, the company does not have many established products and services nor does the market believe it will be able to innovate and create new ones. If this company has any tangible assets, it should try to sell them

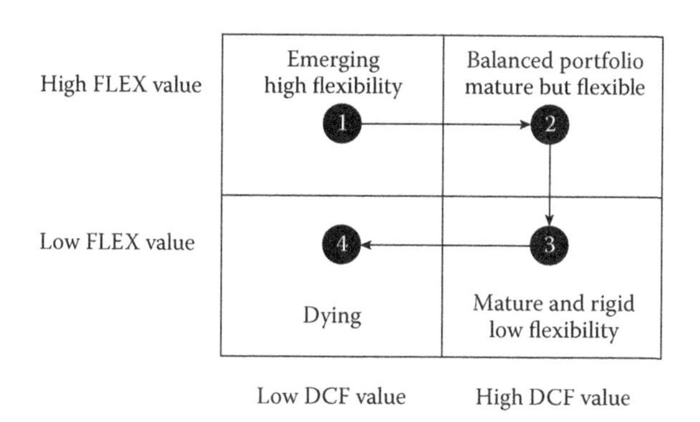

FIGURE 7.5
Life cycle pattern of large companies over a long period of time.

and return the proceeds to shareholders who could deploy them in more productive ventures.

There are many historical patterns we can identify in this schema of DCF and FLEX value. Figure 7.5 shows a pattern that is followed by many contemporary large companies.

In stage 1, a young and emerging company was formed by visionary entrepreneurs. The company was highly flexible and innovative. The market liked the company and was willing to provide a premium and assign a high FLEX value even though the company had no or few products and services. Over time, the company did not disappoint the investors and it created breakthrough products and services with high market acceptance. Thus in stage 2, its revenues and profits soared. The stock price of the company moved higher even faster as the market anticipated bigger and better things to come from the company. At stage 2, the company had a diversified and balanced portfolio of products and services and investors could not be happier. The company continued to retain flexibility in many aspects and that portends well for future innovation. But a few years later, the story changed. Success sometimes has a tendency to breed arrogance and the company began to make decisions that drained flexibility. They brought in some big guns to grow the company fast and scale it up. But like the commissioning of a new pyramid, the new leaders started with grand plans. They hired indiscriminately and went on a shopping spree, buying other companies wherever they could find them. As scale and higher revenues were the targets, the goal was to get as big as possible and as fast as possible. In board meetings, the pyramidal leaders were criticized at their inability to grow faster than they already had been. As the revenues of the company soared, the executive compensation and bonuses also soared. The compensation leverage (the ratio of the highest to the lowest compensation) in the company began to show rapid rise. They created many different human structure layers and the pyramid was taking form. Acquired companies and organic growth created many disconnected divisions and subdivisions inside the company. Divisions embarked on large technology implementation projects on their own and they competed against each other on how much money each could spend. Planning and strategy departments drew out plans to increase the company's infrastructure, products, and employees. However, as the company grew and increased revenues, cracks were forming at the core. The company had no defined culture and the ad hoc mixing of many different acquired companies created a soup that was going bad. As the leaders of the company were laying out even grander plans in analysts' meetings, the company's stock price began to fall. The company reached stage 3, in which it still had robust profits from established products and services (high DCF value) but low FLEX value. In spite of the assertions that investors did not know the company well, the market began to reduce the FLEX premium it was assigning the company before. The market was indicating that the company had lost the flexibility to innovate and it did not expect many new products and services from the company in the future. Finally, the company slowly drifted into stage 4 in which its established products slowly became obsolete. Given the low flexibility, the company had no chance to innovate further and it would likely disappear into a case study in business schools of how not to run companies.

Another pattern is that of a company formed with much fanfare and expectations but quickly turns into a dying company. In this case the young company quickly turned into a rigid one unable to innovate and bring any products to market (see Figure 7.6).

This may have been a case of venture capitalists (VCs) who funded the company attempting to make money fast by following a standard template. In this pattern, the VC, who agreed to fund the company, installed a management team of industry veterans "to flog the company into shape." They had to do this quickly so that an "exit" can be reached to make returns in excess of 50%. In the VC contract, they added many features to drain much of the value of

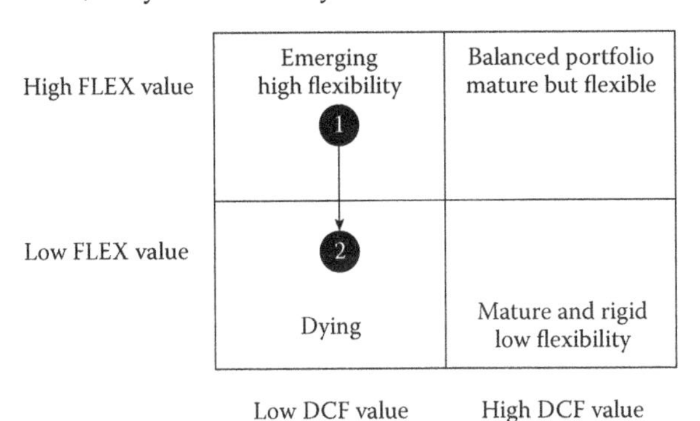

FIGURE 7.6
Pattern of a young company with quick death.

the company from the founders in a well designed "get rich quick scheme." However, for reasons unknown to the VCs (yet), the company began to falter. The people associated with the company lost motivation and the great ideas they had all began to wither away. The company's energy seemed to disappear as it slowly drifted to stage 2 and remained there, waiting to die. Till this day, this remains an enigma to the VCs how such a company with great promise could fail so quickly, in spite of their installing a great management team and providing ample help in operations if not in equity.

A third common pattern is a company starting with products close to market and a great promise of future products. For a variety of reasons, the company is only successful in bringing only the first prototypes to market. After that it is unable to innovate and bring new products and services (see Figure 7.7).

This may be the case of a company unable to grow optimally or being too risk averse. The company milked its first set of products and services. In the process, it took the eye off the pipeline and no further innovation happened. This may be a case of an arrogant entrepreneur trying to do everything himself. He did not want to delegate and wanted to continue to run the company. His technical excellence that spawned the company eventually became a liability as he refused to create a scalable proposition and further innovations at the company. This type of company will slowly die once its marketed products and services become obsolete or loses market share.

Finally, we can also imagine a pattern of catastrophic failure (see Figure 7.8). In this case it appeared that the company is on course to become great. However, it encountered a shock—such as a loss of patent, lawsuit, or fraud.

Although the company had flexibility in some aspects of its business, it was unable to counter the shock and it quickly slipped into oblivion. As we have seen in the preceding chapters, there are many dimensions of flexibility. A company may exhibit good flexibility in certain areas and rigidity in others. For example, if the content systems of the company are not well developed it may be prone to catastrophic failure under a shock.

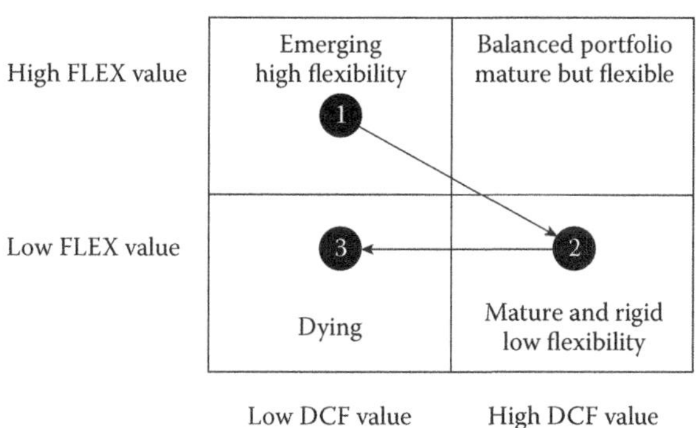

FIGURE 7.7
"Flash in the pan" pattern of company life cycle.

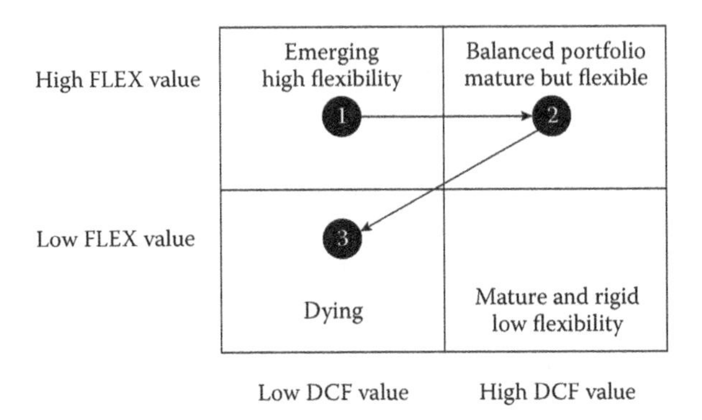

FIGURE 7.8
The pattern of catastrophic failure.

Calculation of FLEX in a company is a complicated exercise. To do so, one has to apply the correct DCF analysis on known products in the market. The difference between the market value of the company and the DCF value is the FLEX value of the company. To make this exercise easier and to get an overall picture, let's use a simple proxy that everybody is familiar with—price/earnings ratio. The price is the stock price of the company and the earnings are the profits/share currently. The current profits/share is a proxy for status quo and the known products in the market. So, we can use profits/share to be part of the DCF value. If the company does nothing, its earnings from marketed products may continue for a few years. They will eventually stop as the company's products become obsolete or become uncompetitive due to changes in the environment.

The other component that affects the DCF value is the *book value* of the company. The book value is an accounting metric but it is a proxy for what the known assets of the company—such as machines and buildings—is worth. It is approximately the price at which the company can sell its machines and buildings if it were to shut down operations today. So the book value coupled with a few years earnings represent the company's DCF value.

If we assume that five years is a reasonable horizon for the earnings from marketed products to continue, we can the calculate the DCF value of the company approximately as

$$B = \text{Book value per share}$$

$$E = \text{Earnings per share}$$

$$DCF = \text{DCF value per share} = B + 5\,E$$

The market price, however, is a combination of the DCF value and FLEX value. So, we can consider earnings to be part of the DCF value and stock price to be the sum of DCF and FLEX value. Thus,

$$\text{FLEX per share} = \text{Price} - \text{DCF}$$

We can also calculate a FLEX % as a ratio of FLEX value to the total value of the company. Mathematically,

$$\text{FLEX \%} = \text{FLEX/Price} = (\text{Price} - \text{DCF})/\text{Price}$$

Figure 7.9 shows the histogram of FLEX % for 4,000 U.S. stocks. On this metric we find 40% of the companies showing no FLEX value. Of the remaining 60%, only about 20% show significant FLEX of more than 100%.

Conventional financial analysts call the high FLEX companies, "growth companies" and low FLEX companies, "value companies." Growth companies thus fall in the upper half and value companies in the lower half of the FLEX-DCF value matrix. Generally, growth companies also show high beta. Flexible companies, however, should not exhibit a high level of beta as the flexibility should allow them to create returns less correlated with the macroeconomy. Conventional categorization of value companies are typically accompanied by low beta. The idea of a value company is that its stock price provides an attractive point of entry. But if the value of the company is largely in the DCF component, there may be a good reason for it. So, the conventional categorization and stock selection by experts may miss the truly flexible or rigid companies. As evidenced by the track record of stock pickers who use deterministic techniques and categorization, it is difficult

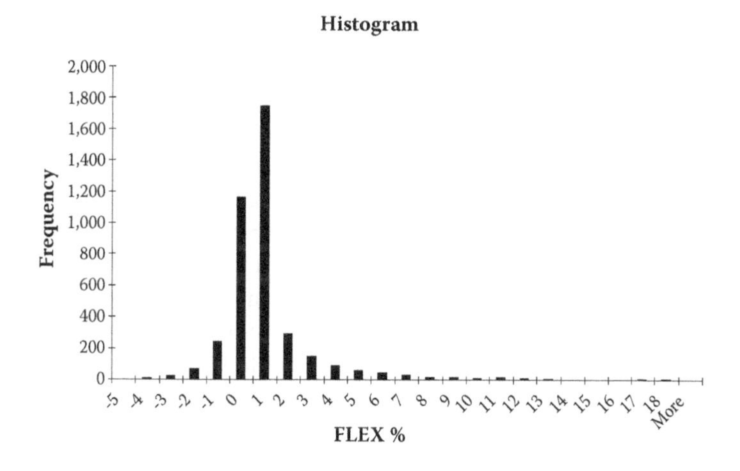

FIGURE 7.9
FLEX % histogram for 4,000 U.S. companies.

to speculate on attractiveness of a company's stock price from known information. It is the unknowns in the company that really determine future stock price movements. However, if we can use recent historical data to get insights into how the company is managed (whether or not it has flexibility), it is likely to be much more predictive of the future success of the company and its stock price.

Brief History of a Technology Company Losing FLEX

Mergers and acquisitions (M&A) are classical signs of weakness and managers' attempts at buying time, hoping that the situation will improve. For example, consider a technology company, PureTech, facing an eroding innovation rate due to flexibility problems in its structure and systems. Although it has many successful products in the market, its R&D continues to fail to produce exciting new products. A decade ago, the value of this company was largely driven by the FLEX component, representing 70% of the total value. Since then, although the company's revenue from marketed products showed growth and the fundamental financial metrics such as growth and margins appeared healthy, the company's market value began to decline. Financial analysts argued over and over again that the company's stock was a great buy. As the stock price declined, many argued that the decline in stock price presented an even better opportunity for investors to profit from the low price. They argued that the market did not understand how well the company's marketed products are performing. For some unknown reason the market was undervaluing the company. In analysts' meetings, the company's senior managers presented compelling arguments as to why everybody should buy its stock. In spite of all this optimism, the stock price continued to decline. Figure 7.10 shows the progression of three major metrics: sales/year, stock price, and percentage of the stock price represented by the FLEX component (FLEX %) from 10 years ago till about 5.5 years ago.

As the company's revenue climbed and its stock price fell, the FLEX component of value in the stock price fell sharply. This indicated that there were many issues within the company that reduced its ability to innovate and create future value. The reaction of the management to this scenario was the acquisition of another large company. This instantly increased overall sales of the company. Counter to the arguments made for the acquisition by the management and faithful analysts, the stock price took a dive and continued to fall. Figure 7.11 shows the representation of the sales, stock price, and FLEX component of value from about 5.5 years ago till 2 years ago.

The spike in revenue at 5.5 years is due to the large acquisition conducted by the company. The acquisition did help the company increase aggregate sales

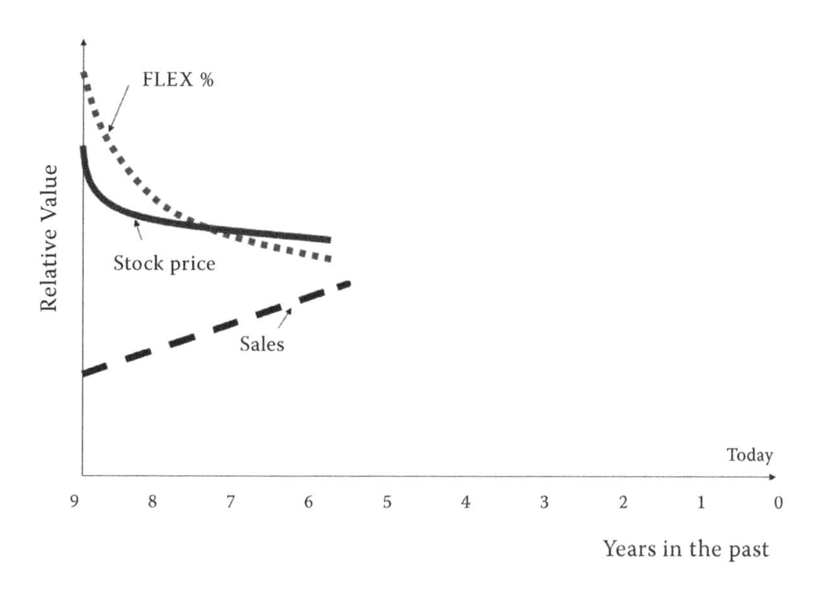

FIGURE 7.10
PureTech's sales, stock price, and FLEX % from 10 years ago to 5.5 years ago.

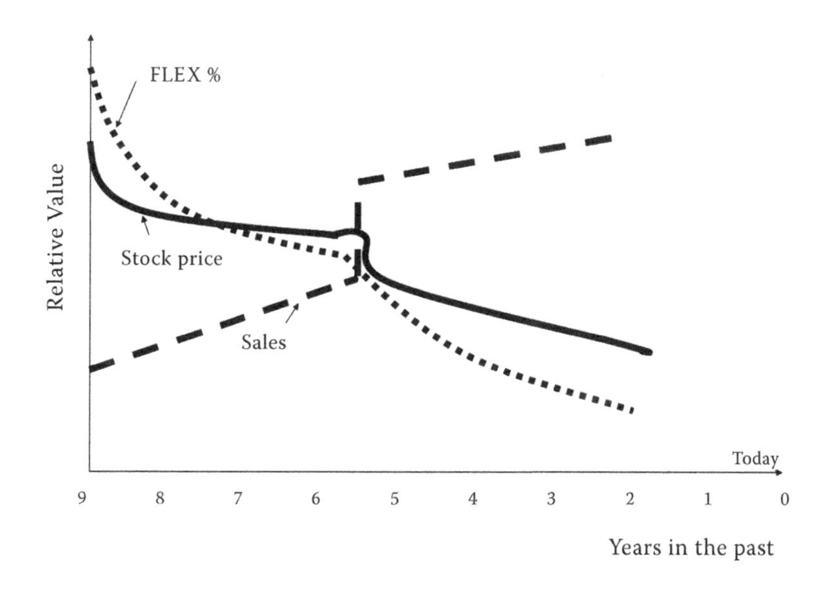

FIGURE 7.11
PureTech's sales, stock price, and FLEX % from 10 years ago to 2 years ago.

and profits but it failed to enhance the value of the company as represented in the stock price.

Since M&A has been the favorite strategy for the leaders of traditional companies, there are many advisory companies specializing in this area. There is a well-defined process to conduct M&A. All the CEO has to do is to signal the intent to acquire. Investment bankers will scour the industry, indentify the right targets, perform financial analyses to demonstrate accretion to sales (if not profits), and make the transaction happen using extremely efficient processes. After that, consulting firms will help the leaders of the acquirer to fully integrate the two companies using copies of templates from many such acquisitions in the past. The M&A process, thus, is a highly defined and efficient process today and will not disappoint the target shareholders, managers, investment bankers, and consultants. However, shareholders of the acquiring company should be asking whether acquisitions are for increasing shareholder value or for increasing aggregate sales so that the managers of the company can command higher salaries. A curious and little known fact is that revenue and total number of employees in the company is correlated with the compensation of executives. The thinking is that larger companies are more complex and managing such companies requires a special breed of people. The market prices of executives who have managed large companies are higher than those who run smaller or medium sized companies. This provides incentives for executives to increase the scale and revenue of companies regardless of the impact on shareholder value.

As the stock price of the technology company continued down, the next reaction of the managers of the company was to make yet another acquisition. Figure 7.12 shows the predictable reaction in the stock price and FLEX value percentage.

The most recent rumor in the company is that the management will attempt a merger in an effort to save the company and increase shareholder value. Unfortunately, this movie has been played over and over again in many failing companies with the same disastrous results. Loss in the value of a company and especially a decrease in the FLEX component of value indicate problems in the structure, systems, and strategies pursued by the company. This cannot be fixed by buying another company. Such a strategy is a diversion technique by managers who would rather increase the size of the firm rather than its value for job security and the elevation of their own status.

It is possible to imagine an event horizon—a zone of no return for companies pursuing such policies (see Figure 7.13). This event horizon is a function of FLEX % and if it falls below a certain threshold value, it may be impossible to introduce sufficient flexibility to resurrect the company.

Once the company is through this critical juncture, the only way out is closing it down or spinning off components as separate companies. Incremental adjustments will not be able to save the company. In the story of PureTech, the event horizon was the first acquisition. When the managers of the company

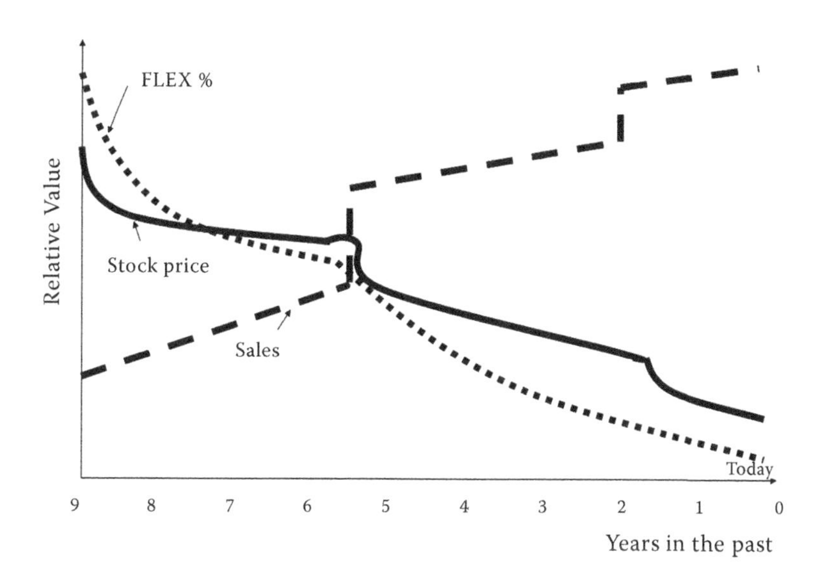

FIGURE 7.12
PureTech's sales, stock price, and FLEX % from 10 years ago till now.

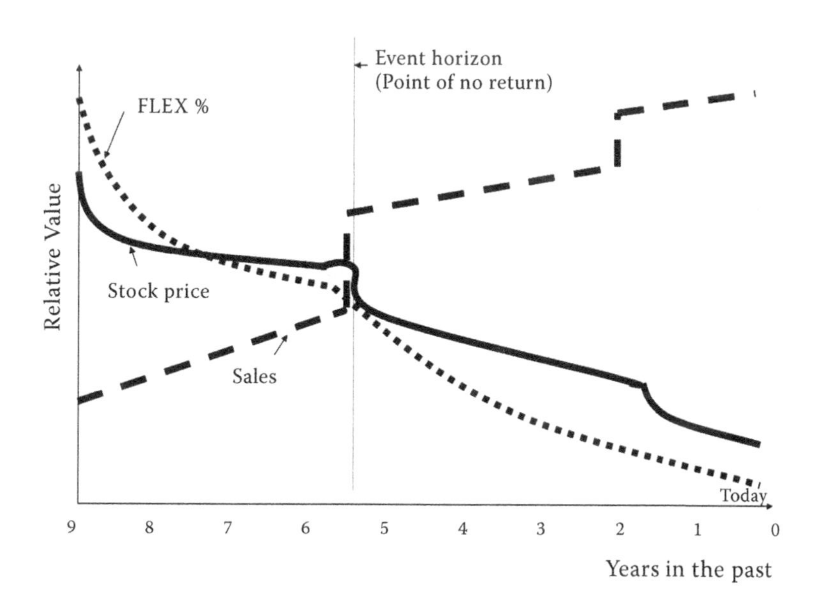

FIGURE 7.13
The event horizon for a company in a downward spiral.

decided to acquire another company so as to fix its internal problems, they demonstrated a level of incompetence that destroyed the company. The market and the shareholders were well aware of this and the fall of the stock price immediately after is validation for it, in spite of the positive comments of the analysts and assurance of the management.

Measuring Flexibility

We can measure flexibility in the primary components of the company—structure, system and strategy as well as all of their sub-components. Figure 7.14 shows the nine primary subcomponents described previously (see Chapter 3).

The flexibility score of each of these components can then be combined to create an overall score for the company. Just like a credit rating that measures financial health of traditional companies, the flexibility rating provides an overall assessment of the company's ability to survive and succeed in an uncertain environment. Once a company identifies problem areas, it can take actions to improve flexibility in the targeted areas.

Metrics for Human Structure Flexibility

There are some telltale signs of human structure flexibility problems in traditional companies. Some of these are explicit signals. For example, low diversity in human resources in terms of gender, age, race, location, and other attributes signals flexibility issues in the human structure. A complex and layered organizational structure, similarly, imply low flexibility. There are many subtle signs also, such as concentration of attributes. For example, gender concentration is common in higher layers of pyramidal structures. Concentration happens due to hiring and retention policies of the company

Infrastructure	Information	Human
Technology	Process	Content
Internal	Boundary	External

FIGURE 7.14
Components of an organization.

that tend to replicate existing attributes rather than seek diversity. Over many years, such a policy creates high concentration and diminishes flexibility in human structure. The flexibility in the human resources originates from the selection and incentive policies for the employees (members) of the company. Layers in the organization, concentration, and composition of the layers and incentive policies, thus, are factors that allow us to establish the flexibility in human structure of companies. In an integrated world economy, the importance of countries and race are becoming less relevant. Gender and age are more relevant diversity factors and so I will use these two attributes in the measurement of concentration. In a conventional pyramidal company, concentration in a higher layer has a more deleterious effect on the prospects of the company. This is because of the criticality of the decisions made in the upper layers of the organization.

Gender concentration is important because of the genetic aspects of the decision-making processes followed by males and females. From the period humans organized themselves to form a family, the diversity in the thought processes and abilities of both genders helped them to survive and succeed. There was a natural specialization with men with stronger physique taking to hunting and women managing the overall process. This specialization has given both men and women differing perspectives and methodologies to identify and solve problems. If gender concentration exists, companies will lose the flexibility to combine different ideas and perspectives to manage uncertainty. For example, recent studies show that females are better at consensus building and thus may be better equipped to lead large and complex organizations. However, gender concentration is skewed toward males in the upper layers of large companies, making them more vulnerable to shocks.

Age is important because of the diversity brought through experience as well as new ideas in a fast changing world. It is the ability to combine new ideas with experiential knowledge that makes companies succeed. Companies in many countries do not hire below a certain age for certain kinds of jobs. The premise here is that with age comes experience and maturity and the ability to function better in complex jobs. During the Internet era, this concept was challenged and many companies run by young entrepreneurs surfaced. They were no better or worse than many companies run traditionally by experienced managers. The idea of experience leading to improved performance has to be challenged in a world that has high uncertainty. Just as in performance of financial assets and funds, past performance is no guarantee of future performance. What is important is to have a blend of people—some experienced and some not—so that the company can enhance its flexibility in the human structure dimension.

A flexible company will be organized as a SOUL (self-organizing uni-layer) with selection of members through subscription and democratic choices of existing members. This type of company will be diverse in all attributes such as gender, age, race, education, and so on, as these attributes are not used in the selection process (as they are typically in traditional pyramidal structures). Existing SOUL members also have an incentive to increase diversity in all attributes to increase flexibility. A very high percentage of the compensation of the members of the SOUL will be based on accomplishing the end objective of the SOUL as well.

Human resource (HR) flexibility is inversely proportional to the number of layers and concentration in the layers. Modern companies have substantially more layers than companies in the Industrial Revolution. Driven by complexity in the business and specialization, the number of layers in companies continues to increase. This is also affected by the implementation of certain rules suggested by HR experts such as the maximum number of immediate reports and the effective management span. This is based on the idea that a manager can only effectively manage a certain number of people at any time. As is obvious, the smaller this number is, the higher the number of layers in companies. This, however, assumes the general notions of management and human structure. In a flexible company where incentives are well aligned with all members, no such management is needed and the management span can be infinite. Traditional management exists only because of misaligned incentives.

Also, a company will have low compensation leverage (ratio of highest to lowest compensation) if the compensation is based on value added (and not on titles). In a SOUL, every member is critical to the success and no single member is responsible for integrating disparate functions as done by the CEO in a pyramidal structure. As such, it will show a more uniform compensation structure. The compensation leverage, thus, is a proxy for both the flatness and flexibility in a company. A high ratio also signals a lack of information flow, forcing shareholders to focus pay at the top and measure performance there. This is based on the misguided belief that a few people can make a big impact in the management and success of the company.

In the United States, the compensation leverage in large companies continues to rise. Some measures indicate that this ratio rose from 40:1 in 1980 to 400:1 recently. In a traditional company, the CEO or the top manager integrates the various specialized functions and makes decisions that affect the company as a whole. These decisions are indeed critical and they have a big impact on the survival and success of the company. From the perspective of the owners (shareholders), the CEO position and the person who is in that position are too important and hence need to be compensated appropriately. Generally, the compensation of the top executives is described in the context of providing incentives. However, owners use this compensation more for risk management and less for incentives. In a pyramidal structure the loss of the top executive can be disastrous for the company in the short run. Just as the clan leaders harbored proprietary and secret information for the control of the clan, the CEO of the company does this as well. Since the apex of the pyramid is narrow, the company's secrets reside with a single person or a few people. Thus, the loss of the top executives has different implications for the owners. First, the company may get effectively paralyzed as it has to find somebody else to step into the position quickly to run it. Second and more importantly, the information held by the executive could be used by the company's competitors to attack its products and services. The compensation for the CEO and top managers, thus, is a costly insurance policy for the owners. In addition to the positive reinforcement of a high compensation, owners tend to also apply negative reinforcements such as golden handcuffs—constructs that make it much more attractive for the executives to continue on the job than to leave.

In the lowest levels of the pyramidal organization, compensation is used primarily as a contractual obligation. The company hires somebody to do a specific job and that person is paid a specific compensation for that. This compensation is typically fixed. Since the employee's compensation is not related to how the company benefits from the work, no benefits accrue to the employee from the success of the company. From the company's perspective, the wages part of the contract is tied to the job. The company can easily replace the employee at a later time with another person, who will provide the same services to the company. The company (and the owners) thus does not have any concern related to the departure of the employee and they do not need any insurance against the employee's departure.

The compensation of the CEO and the lowest-paid worker in a company are, thus, driven by two very different requirements. The fact that the ratio of the highest to lowest compensation has been rising is an indication that the pyramidal structure is getting more reinforced in traditional companies. Recently, it has taken a more alarming turn in that the boards of companies rubberstamp decisions and compensation plans suggested by the executives. Although the boards are supposed to represent the owners, the concentration of the company's shares with a relatively small number of mutual fund managers means that selection of the board is largely driven by a few people. If the few fund managers have incentives different from the rest of the owners, they may make board selection decisions that are not fully aligned with all the owners. This "governance problem" has become extremely severe in many traditional companies.

So, the compensation leverage in a company reflects how pyramidal the company is and thus is a good proxy for the lack of human structure flexibility. If a company finds this ratio is climbing over time, it also implies a declining flexibility. Since the CEO insurance may be based on the "market rates," it is likely that companies in similar industries or locations will drift together in this dimension. If the active owners (fund managers) are the same in different companies, they will also show similar board structures and behaviors. Recent evidence from the United States indicates all of these trends are true and that companies continue to lose human structure flexibility and their ability to deal with uncertainty. Failure of 100-year-old firms in the recent financial crisis is a sober reminder that inflexibility is fatal under uncertainty. The executives in the failed companies were responsible for making decisions and managing risk in an unfamiliar environment.

The lack of information transparency and segmentation also meant that the decision makers were unaware of what was happening in their sprawling fiefdoms. Such spectacular failures could have been avoided if the companies were flatter and designed to have flexibility.

We can use the following metrics to represent human structure flexibility.

1. Layers
2. Age and gender concentration in layers
3. Compensation leverage

The more layers there are in the company, the lower the flexibility. The more age and gender concentration in the layers, the lower the flexibility. Finally, the higher the compensation leverage, the less flexible the company is.

Metrics for Information Structure Flexibility

Companies tend to collect data in a detailed fashion. For example, companies may track time spent on activities by employees. This data was useful in industrial companies to determine productivity of single individuals as well as overall manufacturing plants. Contemporary companies are largely driven by information generated (and not physical goods produced). As such, the amount of time spent in activities is not relevant for productivity. Many computer systems have made it possible to collect and store a large amount of data in organizations. There is a general belief in companies that more data are always good and that data collection should encompass all available data. This has created a lot of noise and has wasted valuable resources and time.

Data are only useful if they help the company make decisions or conduct a routine activity such as payroll and taxes. Routine data and the processing of it can be fully automated and does not require much human interference. One metric for information flexibility is the use of data in decisions. If the company collects large amounts of data but never use it in any decisions, it loses flexibility due to noise and confusion as well as the cost of collecting and storing it. Collected data also tends to spawn analysis for the sake of analyzing the data but this may not aid any decisions in the company. Companies should try to focus efforts to collect and store just the sufficient amount of data to make critical decisions. Companies should also be wary of analyses that do not specifically lead to any decisions.

In the ideal case, an organization will have only one information structure and that reflects the relationship in information collected and used in all parts of the organization. In general, low information use and low integration of systems across the company will lead to high information latency (time elapsed between capture and use). The more data are collected, the more noise exists, and it takes longer to tease out the more relevant information for decisions. Similarly, if the company has disconnected information systems, it will be unable to transport information efficiently from the point of collection to the point of use.

Another important aspect of the information structure is its definition. Traditional information structures define a data element as a single number. For past information, each data element needs to also have a context. The user of the data will require not only just the number but also the context in which the data was generated. Part of this context is uncertainty.

We can use the following metrics to represent information structure flexibility.

1. Information aiding decisions/total information collected
2. Integration of information across the company
3. Information definition including context and uncertainty
4. Information latency (time between capture and use)

Metrics of Infrastructure Flexibility

The infrastructure flexibility can be measured in two major areas—real infrastructure and financial infrastructure. The real infrastructure of a company includes all physical infrastructure including buildings and plants. The financial infrastructure includes its capital structure and financial position. If a company is able to switch from one infrastructure to another quickly as new information arrives, it has a higher level of flexibility. In general, if a company owns its infrastructure, it has less flexibility to switch to another one as the environment changes. So one metric is the total physical infrastructure owned by the company as opposed to leased infrastructure. Leasing alone does not add flexibility if the leases are long-term and the penalty for breaking the lease is severe. The features of the lease and the contracts become important here. Similarly, a company that owns different types of physical infrastructure, differentiated by size, fuel use, location, and other attributes, has a higher level of switching flexibility. Also, a company that has modular machines in its plants may be able to configure them quickly to meet the demand for certain type of products and certain volume of demand. Equipment that aids more flexible manufacturing processes adds flexibility to the company. Modularity thus is an important enabler of real infrastructure flexibility.

Another component of real infrastructure flexibility is operating leverage. If a company has a high component of fixed cost in its total cost structure, it will have high operating leverage and low flexibility. This is because the fixed costs will prevent the company from rapidly changing as the market changes. For example, automobile companies in the United States have high fixed costs related to the pension and medical payments due to the retirees. The companies have to make these payments regardless of the environment and thus it is fixed. The labor contracts these companies have with union workers are also negotiated many years in advance and allow little flexibility. These fixed costs drain flexibility from these companies and they are unable to make the structural changes needed to survive in challenging economic times. If these companies go into bankruptcy, they may be able to renegotiate the contracts they have with the retirees and union workers and introduce flexibility. Total cost of the company is the sum of its fixed costs and variable costs. Lower fixed costs means higher variable costs and vice versa. The more operating leverage a company has, the less flexible it is.

The financial infrastructure flexibility can be analyzed in the long term and short term. In the long term, the capital structure of the company is important. For example, a company with high leverage and debt has lower flexibility than the one with no debt. Debt represents a fixed obligation for the company. In an uncertain environment, a fixed obligation reduces the company's ability to manage. Leverage is created by borrowing money and creating a fixed interest obligation. If the borrowed money can be deployed to create returns more than the interest payments, the company can manufacture very high returns to its shareholders as they are playing with borrowed money. But if the strategy of the company fails or it does not have sufficient returns so as to cover the interest obligation, the company may go bankrupt. The assets of the company are the sum of its debt and equity. The higher the debt, the lower the equity in the capital structure of the company and vice versa. Note that typically long-term debt is used to calculate the leverage. We want to include all debt (long and short term) because in the current environment the presence of either can substantially reduce capital structure flexibility. The higher the debt in the capital structure is, the less its long-term financial flexibility.

Just as the capital structure flexibility affects the company's ability to manage uncertainty, so does its cash position. In the short run, companies can run out of cash or may not have enough resources to fund opportunities presented by uncertainty. It should be noted that the mere presence of cash in the balance sheet, however, does not guarantee flexibility. Unless the company has immediate access to it when needed, such resources do not add flexibility. If the company has a line of credit it can draw from at will, it may suffice to add flexibility to the financial structure. Cash includes accessible cash, marketable securities and lines of credit—anything that allows the company to raise cash and spend it with very short notice.

We have to consider the company's access to cash in the context of its expenses. Expenses include both investments and routine costs of the company. Cash flexibility is critical for the company to pursue the investments it needs to make. Investments represent capital, research and development, marketing, and other type of investments that the company makes every year. The company also needs to cover its routine expenses. Working capital is a crucial consideration for many companies as a cash crunch can simply drive the company out of business even though it has valuable assets and profits. As the readers are aware, cash flow is an important consideration for individuals also.

We can use the following metrics to represent infrastructure flexibility.

1. Owned versus leased physical infrastructure
2. Switchability and modularity in physical infrastructure
3. Operating leverage (fixed cost/total cost)
4. Financial leverage (total debt/total assets)
5. Cash available/total expenses per year

Metrics for Technology Systems Flexibility

Contemporary information technology systems are created in a disconnected fashion—they collect and store large amounts of data that never aid in making decisions and report data without context and existing uncertainty and do not typically perform in real time. The information technology flexibility is linked to the ability to support the flexible information structure. If technology systems are aiding the four aspects of information structure—use, scope, timing, and context—they can be considered flexible. Thus, flexible technology systems have to be built in such a fashion that they are integrated across the enterprise (scope); collect, analyze, and report just the most relevant data for decisions (use); provide the context and uncertainty in the data (context); and are able to surface information in real time (timing).

Flexible technology systems, then, will not be stand-alone but will be distributed across the enterprise using universal mechanisms for the collection and reporting of the data (such as the Internet) and may store information in a form that is not specific to a department or company. Cloud computing platforms that provide a universal backbone for information and analytics as well as storage outside the boundaries of the physical location of the company may provide much higher flexibility than traditional systems. Many companies are still reluctant to introduce nonproprietary networks in their businesses for fear of service disruption and loss of critical information. This fear, however, is misplaced and may prevent companies from embracing flexible and emerging technologies.

The use of nonproprietary and widely available technology systems introduces a higher level of flexibility for two reasons. First, such systems are used by a variety of companies and industries and will likely change faster as new information becomes available. Because they are used by different types of industries, the vendors of these systems will become aware of economy-wide trends faster and can adapt the systems to take advantage of them. Second, proprietary systems require experts in the company to maintain and change them. In an uncertain world, such systems may become irrelevant or may need fast changes and a company that is not in the information technology

(IT) arena may find it difficult to attract and retain technologists with high competence and ability.

Technology systems also have to integrate all systems including information, manufacturing, building, transportation, and others. Since manufacturing, storage, and transportation of physical goods and the production and provision of services are driven by information, a fully integrated system will be necessary to optimally operate any company.

We can use the following metrics to represent technology systems flexibility.

1. Integration (across the company, platforms, and functions)
2. Generalization (use of industry-wide standards)

Metrics for Process Systems Flexibility

In the 1990s process innovation helped many companies take a quantum leap in productivity and profitability. Supply chains became a strategic aspect of some companies—how they manufactured, stored, and shipped products was equally important as the designs of the products themselves. Supply chain optimization programs allowed companies to become lean and mean, reduce inventory, and institute just-in-time policies. As companies focused on process more, they may have inadvertently taken their eyes off products. As they made the processes more and more efficient, the business model became finely tuned to execute status quo extremely well. Although this is a good policy in a stable regime, it may not be the best in a world with high uncertainty.

Product innovation took a backseat to process improvement in the last couple of decades and that may have reduced flexibility in many industries. Manufacturing products at a lower cost using an optimized process, although good tactically, does not guarantee success for any company. A company has to introduce sufficient flexibility into its processes so that changes, as required by new products and concepts, can be easily implemented. The primary attribute of a process has to be its ability to change fast, rather than the ability to run specified activities in a more efficient fashion. It is not process efficiency that is important in an uncertain world, but rather process flexibility. A process can be considered flexible if it can change quickly to adapt to product changes or it has redundant and contingent branches that allow reconfiguration. Flexible processes (not efficient ones) add flexibility to a company. Adaptability and reconfigurability are the fundamental metrics for process flexibility.

Metrics for Content Systems Flexibility

The flexibility in content systems—cultural, moral, and legal—is difficult to measure. A company with a strong content system will also have a strong culture and a unified belief system. Employees of this type of company are unlikely to be confused about a course of action when presented with unknown and unexpected information. There is an integrated framework that will always provide guidance as to the best course of action. This is not written down in big manuals but is shared qualitatively across the company. One measurement is how many people in the organization articulate what the company stands for and what it considers to be important. If there are fundamental principles that are applied across the company consistently, then all associated with the company will be able to articulate it. This is not just words as conventional vision and mission statements tend to be but rather what really defines the company. If the company does not have such a belief system (that is part of the company's culture), its content flexibility will be low. If the belief system is well established and prevalent, then many employees will be able to define it. Large and bureaucratic companies may be governed by rules and regulations but may not have strong content systems.

Metrics for Strategy Flexibility

The differentiating aspect of the internal, boundary, and external strategies pursued by the company is the systematic and explicit consideration of uncertainty and flexibility in the formulation and execution of strategies. Traditional internal strategies, driven by efficiency and optimization typically lead to rigid outcomes for the company. These include how the company is organized, how products and services are designed and produced, how R&D is conducted, how manufacturing plants are managed, and so on. In most cases, the company collects and analyzes large amounts of static data and concludes on the best way to manage. If a company uses uncertainty and flexibility systematically in the design and implementation of internal strategies, it will have high internal flexibility.

The boundary strategies describe how the company deals with its suppliers, buyers, partners, and collaborators. This is the immediate extension of the strategies the company creates internally. In some cases it is the extension of its supply chain. The contracts it enters with counterparties are of great importance to the success of the company. By embedding flexibility in these contracts, companies can better manage overall uncertainty. By designing contingent flexibility in licensing transactions and valuing it properly, companies can extract higher value for shareholders. If a company designs and implements flexible strategies that encompass its partners and collaborators, it will have high boundary flexibility.

The external strategies are crafted to interact with the external environment—competitors, investors, and regulators. Any strategy based on fixed forecasts of the future are bound to fail. It is not the precise definition of the future characteristics of economies, markets, competitors, products, and services that are of importance in strategy formulation. Rather, it is the design of contingent actions the company can take as it learns more in the future. This can only be done by an explicit consideration of uncertainty and flexibility in all analyses conducted to aid decision making.

In the next chapter I will discuss how companies can use these metrics to diagnose flexibility problems and take tactical actions to improve them. There are many short-term actions that companies can take to improve their current posture and to avoid catastrophic future failures.

8

Diagnostic Kit and Tactics

We now have a set of metrics that can be used to measure existing flexibility in organizations in all three dimensions—structure, systems, and strategy. Figure 8.1 is a schematic that summarizes the metrics derived in the previous chapter.

As demonstrated before, there are nine flexibility metrics that are critical to the performance of any organization. They are equally important and problems in one area could create issues in others. Let's conduct a flexibility audit for a hypothetical technology company. The company, PureTech, has been in business for over 100 years. Its products are known all over the world, and it conducts business in over 30 different countries. It has over 100,000 employees and many R&D centers around the world as well as country-specific marketing organizations.

Let's start with human structure flexibility. The company is organized traditionally with about 20 different layers between the CEO and the plant operator. The company is also functionally organized with R&D, manufacturing, and marketing in different locations and operating nearly autonomously. Each layer of the company shows significant gender and age concentration with higher layers showing higher levels of concentration. The

Infrastructure Switching ability Modularity Operating leverage Financial leverage Cash access	Information Use Scope Context Latency	Human Layers Concentration Compensation leverage
Technology Integration Generalization	Process Adaptability Ease of change	Content Well defined principles Strong culture Less prescriptive rules
Internal Systematic consideration of uncertainty and flexibility in internal strategies	Boundary Systematic consideration of uncertainty and flexibility in boundary strategies	External Systematic consideration of uncertainty and flexibility in external strategies

FIGURE 8.1
Metrics of organizational flexibility.

gender concentration was the highest in the topmost layer with only one female executive in the senior management ranks and the board. Age concentration also was extremely high in the top layers. The pay scale shows high dispersion with the ratio between the high to low salary showing a compensation leverage of nearly 300:1. All these factors point to existing low human structure flexibility at PureTech. This is the case with most traditionally organized companies that have been around for a while. The conventional human resource processes tend to create many layers, increase age and gender concentration over time, and result in a skewed pay structure.

Although PureTech is in the technology business and is information rich, it has traditionally not collected a lot of data. This is partly because of lack of systems and partly because of the high uncertainty it faced in many dimensions. Ironically, this actually provides the company with some information flexibility, as it does not have many bad habits of collecting irrelevant data. Most of the data collected by the company is never used in the decisions made by it, however. Most decisions are made on gut-feel and by qualitative consensus. PureTech managers are proud of their ability to make the right calls. They point to the long and successful history of the company as evidence for it.

Each department in the company collects data for its own use, and they share very little with each other. The company thus has a large number of disparate and disconnected information structures. Only the most conventional structures such as payroll are integrated across the company. Most data collection mechanisms take many weeks before they report data for decisions. The context of the data including uncertainty is never collected and analyzed. All of these point to low information structure flexibility at PureTech.

The company is organized functionally. All functions—R&D, manufacturing, and marketing—have their own systems and many of them are created internally and utilize proprietary protocols. Systems thus are unable to interact with each other and provide a holistic view. They are also unable to change because of their proprietary nature. Because of the hesitation of the company to partner with others, it has built up a huge information technology department internally. Currently the IT department represents as much as 20% of the cost of the company, and since information technology is not a core competency for PureTech, it has not been able to attract talent in this area. So the proprietary systems created by internal IT are inferior to what is available commercially. The company also does not have many real-time systems except in pockets such as manufacturing. All these factors provide a very low score for technology flexibility for the company.

The company has created its own real infrastructure. It owns corporate offices in the center of a large city. Its R&D organizations also created their own plants and other infrastructure. As a policy, the company does not lease much of anything. It does outsource some components of manufacturing and that may allow it to switch some providers at a later time. However, its switching flexibility in the real infrastructure is small. It also created manufacturing plants with large capacities that are not reconfigurable. The plant designs were fixed as the company expected to run large volumes of products for a long period into the future. In early R&D operations, a few small-scale plants were designed to be reconfigurable. However, the modularity-related flexibility in the overall infrastructure is also small. The company's fixed cost compared to its total cost base is moderate, primarily because a high percentage of its expenses are people related and the company has no unions or noncancellable employment contracts. Only about 25% of its expenses can be considered fixed and thus it has low operating leverage. However, in view of its low switching ability and modularity, we can conclude that the company has low real infrastructure flexibility.

The company has access to cash and marketable securities in its balance sheet. The total access to cash and credit is two times its yearly expense-pool, giving it an apparent high cash access. However, 50% of the cash is abroad and if that cash is brought to the United States, where the company is headquartered, it will incur high tax penalties. Effectively, thus, its cash access is only half of what is evident in the financial statements. It also has very low debt in the capital structure, giving it low financial leverage. The low financial leverage combined with high access to cash gives the company high financial flexibility. Although the company has low real flexibility, its high financial flexibility coupled with low operating leverage provides it with moderately good infrastructure flexibility.

PureTech is fundamentally an R&D-driven company as a large percentage of its revenues comes from new innovations. Its processes thus are targeted at innovation. The company never really focused on efficiency and process optimization although there may be some areas in the company that could benefit from it. Its processes, however, are not designed to be flexible. So, in some sense it has the worst of both worlds. The company targets products with high market potential and has designed processes—R&D, manufacturing, and marketing—to be aligned with the high revenue expectations from the end product. These processes are not flexible enough to accommodate other types of products and that has resulted in low process flexibility for the company. Certain processes in the company, such as raw material procurement and early research and development, are flexible as they are able to adjust to different types of products. However, the economic value of the output of these processes is low compared to the total output of the company. Overall, this gives the company low process flexibility.

The company grew through acquisitions. In its current incarnation, it has incorporated at least three other large companies. This has created great confusion in the culture of the company as it struggles to define an overall set of principles for the company to operate on. Since the employees of the company came from different organizations, they are still trying to understand the differing jargons. Additionally, the company's legal and ethical systems are based on highly prescriptive rules. The philosophy followed by the company in acquisitions was to fully convert the target company according to the set specifications. The company has been in the habit of using consulting companies to integrate the acquired companies according to industry standard templates. This has resulted in the company losing valuable information and business practices that existed in the acquired companies. Overall, both the large acquisitions conducted by the company as well as the integration practices followed by it have resulted in an organization with low content flexibility.

Certain parts of the company, such as R&D, had to deal with technical uncertainty routinely. Technical uncertainty is related to prototype failures due to unexpected technical data. As such, the strategies pursued by R&D always considered technical uncertainty. However, they seldom considered market uncertainty explicitly in decisions as much of the R&D has been disconnected from marketing. R&D strategists are also keenly aware of the various options and contingency plans they can pursue. However, they lacked methodologies and tools to systematically incorporate uncertainty and flexibility into decisions. Since the focus has been on technical uncertainty, the company has routinely used decision trees and other approaches that allow the capture and representation of technical uncertainty. Ironically, this has led the company to wrong decisions in cases where market uncertainty and decision flexibility are important considerations.

In manufacturing, the objective function for optimization also typically assumed a certain time window, prototype success, and marketing scale. In early R&D, there has been a higher level of appreciation for overall uncertainty—both market-based and technical—and there have been efforts to incorporate them into decision making. However, in most cases, decision making is done in a stage gate process with gates many years apart, allowing little real-time feedback and incorporation of available new information. There have been many issues related to the design of incentives because of the lack of feedback in a long R&D process. As such, politics and segmented incentives played a dominant role in crafting internal strategies in the company. Overall, the company has low internal strategy flexibility.

Although the company is vertically integrated it does have partners and collaborators, especially in R&D. It has been engaging with outside companies in many dimensions—innovation, information sharing, technology, and manufacturing. In many of these transactions, the company had to explicitly consider technical uncertainty. There are also operational uncertainties related to coordination and regulatory interactions. The company had to design in flexibility to manage some of this uncertainty in contracts and licensing arrangements with partners. We will consider the boundary flexibility to be medium for the company.

The external strategies followed by the company assumed a constant throughput of innovative products from R&D. As such, the marketing organization was scaled up and it managed all sales and channel activities internally. This also led the company into collaborative deals with other companies, driven by its insatiable need for new products that were not forthcoming. The company's future plans and strategies have been static based on expectations. There have been few contingent plans and decision flexibility seldom figured in strategy design. The company has been reliant on advisory firms for industry and competitive analyses. Most of these types of information aggregated trends and forecasts based on what is known and can be forecasted as a trend. Rigid strategies based on constant forecasts coupled with a scaled up internal organization with expectation of new product throughput drained most of the external flexibility from the company. We will rate the company low in this dimension.

Figure 8.2 is the representation of the results of the flexibility diagnostics of the company. The diagnostic kit shows problems in many aspects of the company. Except for the infrastructure flexibility (driven by low debt, moderate access to cash and low operating leverage) and moderate progress in utilizing flexibly in contracts with partners and collaborators, the company harbors problems in every other area.

Let's look at the tactics PureTech can deploy to improve overall flexibility. The first area to focus on is the human structure. Here the company has to make some dramatic changes. First, it has to fundamentally de-layer the company. For this company with 20 layers currently, halving it to 10 within one year should be a reasonable goal. Two years after that, the company should try to halve layers again and reach a more reasonable five-layer structure.

Infrastructure Low switching ability Low modularity Low operating leverage Low financial leverage Moderate cash access	Information Low information use Low information scope High information latency No information context	Human Large number of layers High concentration High compensation leverage
Technology Highly proprietary systems Low integration	Process Low adaptability Not changeable	Content Low definition in culture Prescriptive rules
Internal Low	Boundary Medium	External Low

FIGURE 8.2
Summary of the flexibility audit for PureTech, a hypothetical technology company.

How fast PureTech can change will depend very much on the desire of the top executives to make it happen.

De-layering can happen in two directions. Top-down, it can take out intermediate layers that do not substantially add to decision making. Many of the middle management layers are used both to buffet the layers below as well as a punching bag for layers above as the situation demands. Traditional thinking has been that the "management-span" cannot exceed a certain number of subordinates. This is because there are only so many people a single person can manage. The fundamental question here is why people need such close "management." If the layer below the CEO requires "management," the company has to first fix its incentive systems so that people are motivated to do the right thing and do not require heavy-handed management. If this can be done, the "management-span" can be increased manyfold and many of the layers eliminated. This process can flow top-down reducing the top 10 layers to 5.

Bottom-up, the company can combine layers and de-layer that way. One curious phenomenon seen in traditional companies is the existence of a large number of titles. Titles are sometimes used as a reward mechanism as the employee compensation is invariably tied to them. Over time, these titles tend to accumulate and create many layers and sublayers in the company, destroying its structural flexibility. One way to de-layer bottom up is to eliminate traditional titles. For example the chain—associate, manager, and senior manager—naturally fit in and aid layering in a company. If the company moves into a structure with no titles or titles based on specialization—such as information architect, process designer, compensation consultant, and so on—it will be in a much better position to de-layer. Some forward-looking technology companies have allowed employees to create their own titles to inform others what they are good at and what they are passionate about. In this case, titles will not create natural layers in the company.

Another important aspect of de-layering bottom-up is providing employees with the ability to define their own jobs and schedules. In a conventional company, "an associate" may have a set of defined job requirements including location and timing. The compensation of the associate may be fixed by the scale created by the human resources department. In many cases, people interested in certain aspects of the job may not be able to take it because of the constraints around it. For example, a mother with children attending primary school may want to have less pay but more location and timing flexibility. In a layered and traditional company, this may not be possible. By making job descriptions less defined and more customizable, PureTech can reduce the number of bottom layers.

PureTech also shows high concentration of age and gender at all levels and especially in top layers. This is symptomatic of the human resource procedures currently in place that allows existing participants in layers to replicate themselves more easily. Alleviating concentration once a company reaches a certain level of rigidity is tough to do. Any policy that "requires" hiring managers to reduce concentration will result in suboptimal hiring decisions. So, it will be unwise for the company to try to solve one problem by creating another. One way to reduce concentration is to remove the traditional hiring practices and job descriptions. If the company can incorporate more objectives- and subscription-based human structure development, it can slowly unwind the shackles of concentration. For example, the company may want to rotate employees across layers, functions, and jobs. Such a policy may allow nontraditional participants to consider responsibilities in areas that are traditionally closed to them. In doing so, the company also increases flexibility in the work force by building more generalists who have broader understanding of the company's business.

To reduce concentration in the long run, the company has to redesign its hiring and retention policies. By involving a larger number of existing employees from differing levels in the identification and selection of new employees may help the company avoid departmental cloning. Concentration is initiated and perpetuated by departmental managers who attempt to replicate themselves in the hope of reducing conflict and enhancing consensus in decisions. In the process, however, they end up with people who think alike and always agree. The department managers' goal of lesser conflict and more consensus is accomplished at the cost of higher concentration and lower flexibility. By involving a larger number of existing employees from all levels and departments, the company has a higher chance of finding employees that complement and eliminate gaps in its existing competencies.

One way to reduce concentration in upper layers is to allow the selection of top layer participants by a democratic vote. Since the employees of the company (as a group) have information on the company's products and strategies as well as the people who may best lead the company, they are in the best position to hire the top managers of the company. For example, one could conceive a 4-year election cycle in which the employees of the company will elect the C-level executives by electronic vote. Existing executives may lose their jobs in this process if they are not able to increase the value and communicate effectively with the employees. This process can be conducted for most of the layers in the company and not just the top layers. With intranet-based technologies available, such a process will be easy to conduct. If done systematically, the election process will also improve information flow in the company. Readers may be aware of politicians becoming much more enthusiastic about sharing information with their constituents when elections draw near. However, the owners of the company have to ensure that the leaders of the company are not just politicians but good leaders. If the entire company participates in the election process, it is unlikely that they will elect incompetent managers who may run the company to the ground.

PureTech also has high compensation leverage. The ratio of the highest paid to the lowest paid employee is currently very high (300:1) and that portends significant problems. Till recently, the argument for high CEO pay has been that the people in the executive level are a special breed and the market price of that talent is high. As the world witnessed spectacular failures of large institutions in recent years, this argument has lost the punch. Companies run by highly paid executives have destroyed much shareholder value in the past decade. Although they have not been able to add any value to the company, the pay still remains high. Bonuses are paid in good times and bad. In good times, the talent of the executives was considered to be the reason for the company's success. In bad times, the economy was blamed for the company's misfortunes. No single individual, however smart he or she is, contributes 300 times more to the success of the company than the individual at the bottom of the ladder. For PureTech, a complete redesign of incentives and pay is needed. Pay has to be based on value added rather than the position occupied in the pyramid. PureTech has to sever the relationship that currently exists between pay and titles. Pay should also not depend linearly on tenure, age, or experience. The company has to design a performance management system that measures the value added by the members. The company should minimize salary and increase the component of compensation driven by the success of the company. The stock price of the company is a reasonable measure of the value added. However, employee stock options have many deficiencies. Restricted stock may be a better way to administer compensation and this should be equally prevalent in all layers of the company and not just in the top layers. Through a better compensation, incentive, and performance design, PureTech should attempt to decrease the compensation leverage from the current 300:1 to a more acceptable 50:1. This can be largely done by substantially reducing the fixed compensation to the top executives to the company. If the executives are really interested in the success of the company, they will be the first to admit that their own pay is highly skewed in relation to the value added. The company may want to completely eliminate fixed salaries above certain layers.

To improve information flexibility, the company has to make improvements in multiple areas. To do so, it has to start with the decisions the company makes. There are decisions at multiple levels that require information from all aspects of the business. The first task is to create a decision hierarchy. Some decisions are routine and can be mechanized, some occur infrequently but can be anticipated, and some others happen randomly. In a decision hierarchy, decisions are portrayed not by their "importance" or "timing" but rather the type of information used by the decision maker. For example, a plant operator makes routine decisions regarding production and that depends on a variety of standardizable inputs in demand and capacity. This can be contrasted with a decision to expand capacity and such a decision is not routine but can be reasonably anticipated. This decision takes a wider set of information related to the company's production and logistics network, location economics, and manufacturing synergies. Some of this information is historical and others may be forecast. Decisions that are driven by shocks such as a patent infringement lawsuit against the company cannot necessarily be anticipated but can occur. So, the company has to have information that will allow it to make quick decisions in such situations as well.

Once the decision hierarchy is identified, the company can create an information structure to support it. The context of the information needs to be considered in the structure. If the company is using forecasted information, the uncertainty in forecasts has to be part of the structure. If the company is using historical data, it has to be cleansed and stored with context. Part of the context will be the statistical aspects of the data. For example, if the data related to the production of a specific product is part of the information structure, the information structure should contain average, standard deviation, skewness, and other cross-sectional attributes of the data and not just a large amount of raw data. In many cases, the company does not necessarily have to store the raw data and can condense it down to a much smaller information set if context can be included. Another example of context in information is the data stored regarding R&D. If the company has many prototypes in research and development and it has a stage gate process for decisions, the raw data of prototypes progressing past the stage gates, albeit useful, is not enough. The information structure has to include qualitative context such as the reason for failure, delay, abandonment, and other decisions the company has taken in the past. The tendency in scientifically managed companies is to keep a high level of focus on the numerical data and less on the qualitative context. Cross sectional analysis of R&D metrics may not help the company improve decisions without the consideration of the context surrounding the data. The numerical focus can also lead the company down the path of "management by numbers." Companies also routinely conduct "benchmarking" exercises on numerical throughput to compare themselves to industry peers. This is another classic example of wasted effort and the use of irrelevant data. Those engaged in benchmarking should ask themselves what decisions are improved by knowing how good, bad, or average they are compared to others. Benchmarking is a sure way to move the company to mediocrity.

Once the information structure is defined based on the decision hierarchy, the company can focus on latency. Latency is the time elapsed between the capture and use of information within the company. Information latency increases with the amount of information captured and decreases with information scope. The information scope is how holistic and connected the information structure is. For example, the company's current practice of separate and unrelated information structures for different functions such as marketing and R&D means that much of the information will remain hidden from eventual users. If R&D needs to make a decision regarding a prototype, it has to request information from marketing regarding the expected profit characteristics of the eventual product and the uncertainty around forecasts. However, marketing may not have the information structured this way and it may be using different terms and jargon to represent the same information. Or even worse, functions in traditional companies may hoard information to curry favors at an opportune time. Data sharing may also be hampered by the fear of the "wrong use" of the data. This is an important issue in large and complex companies, where functions such as marketing and R&D do not share information because they fear that the receiving party may not understand it or use it properly. Even if marketing understands the request, has the right information, and is willing to share it, valuable time will be lost between the request and arrival of data. This increases information latency and so it is important for the company to have a unified information structure and embark on an education process for all functions about the importance of information sharing. Because the company has had disconnected information structures for a long time, it may not be possible to incrementally improve it, as old habits die hard. The best option for this company is to start over with a blank sheet of paper—identifying a decision hierarchy across the company and defining an information structure that will support it well. Since incorporation of the information context will be a totally new aspect and cannot be back fitted to existing structures, it is best to start over. In doing so, it can actually create and test an alternative to existing information structures in parallel and slowly transition all activities into the new structure over time.

Since the information structure design cuts across the entire company, it may also be able to rationalize its technology systems at the same time. In the technology area, the company is managing a set of disparate proprietary systems unable to connect with each other. This is the result of many years of system development by the information technologists employed by the company. Since IT is a fast moving area with new technologies coming on line virtually every day, the internal IT department has been unable to keep its personnel trained and up to date. This has resulted in the use and deployment of outdated tools and systems. The best way for the company to fix this is to outsource all its technology systems to outside expert providers. The company can enter into measurable performance-based contracts with the IT partner. Outsourcing the design and implementation of technology systems also help the company to be on the cutting edge of technology. Since firms specializing in IT will be in tune with the industry trends and available technology, they will be able to select the best options for the company. The company may need a small internal technology management group who can ensure that the technology systems are designed and implemented with the whole company's information structure in mind and not the needs and requirements of one function or the other. In designing technology systems, the company has to consider the integration of information technology with manufacturing and other physical infrastructure technologies. This integration may allow a centralized decision hub that can make intelligent or directed decisions.

The company's real infrastructure does not have the ability to switch or become modular. This is because of years of overinvestment in traditional manufacturing plants and equipment that are tuned to making large batches of products with long lead times. This has created rigidity in the company and has adversely affected its ability to think more creatively about R&D and manufacturing processes. The network of plants and equipment the company currently owns are spread across the world, giving it some level of location flexibility. However, their designs and capacities are similar. This does not allow the company any options to manufacture different size lots or speed up or slow down the manufacturing process as necessary. Most of the plants work on an all-or-none basis and they cannot be operated in a modular fashion. This is a tough problem to solve. One approach is to retain some of the plants in the network and complement them with newer plants of different sizes, capability, and modularity. This will require big capital investment and the company has to conduct an analysis to determine if such an investment is worthwhile. Another alternative the company has is to increase outsourcing and develop a set of providers with different locations, capabilities, and capacities outside the company. By selecting external providers with differing attributes, it can increase flexibility in its real infrastructure.

It can also enter into creative contracts with suppliers that further increase flexibility. Features in supplier, buyer, and partner contracts can be designed to add flexibility in the supply chain. For example, the company can design a pricing scheme with a decreasing schedule of prices as the overall quantity ordered increases. It can also have certain sequestered capacity (of varying kinds) at the suppliers that may incur a penalty if the company does not use the capacity but allows a price break for the committed capacity. The company can optimize the sequestered capacity and take advantage of the price breaks provided by such a contract feature, as it knows more about its demand patterns. The information transferred to the suppliers allows them to manage their part of the supply chain better. By extending the available information about demand uncertainty across the supply chain, the company can increase overall flexibility in the network.

The company shows high financial flexibility due to its low leverage and access to cash. However, part of the cash of the company is constrained as it resides abroad and the company will incur penalties if it were to repatriate it to the United States, where it is located. If the company would like to take full advantage offered by such flexibility, it should have a plan to migrate cash to a location where it has full access to it without constraints. One way to do that will be to use cash in locations where it is tied up, perhaps through selective investments to increase flexibility. In considering its financial flexibility, the company has to also consider currency effects. For example, if its primary domicile has a depreciating currency, the company may gain value by keeping its cash in other domiciles. However, if the company has undertaken programs that counteract the currency differential, such as currency hedges, it may lose the flexibility it may naturally have by combining the cash flow generation in various currencies and the ability to keep cash in various domiciles.

Given that the company has excess cash abroad, the best strategy will be to utilize that cash or create real asset effects that counteract it. For example, if the company were to contract with another firm in the domicile where it has hoarded cash for tax reasons, it can pay the partner in local currency and drain it over time. Alternately, it can invest the cash in real assets such as new manufacturing capacity to improve overall network flexibility. If it has state of the art capacity it is not fully utilizing, it can then sell such capacity to others. In any case, the dominant strategy for the company is to utilize the hoarded cash abroad so that it does not have any constraints in the use of cash in the future.

The company does not currently have flexible processes that allow it to incorporate a higher level of creativity and increase its ability to manage future shocks. Traditionally, it also has not focused on efficient processes because of the belief that the primary goal of R&D is to create products that get to market as fast as possible. As the company's profits decreased due to lack of R&D productivity and the confusion created by a myriad of acquisitions, focus turned to cost cutting and the creation of efficient processes. In the process, the company drained further flexibility from its R&D operations.

The company should consciously try to design and implement flexible processes that are able to incorporate new information quickly. For example, the R&D manufacturing processes currently employed by the company are static—able to create required R&D materials only in large batch sizes and only with long lead times. All the decision processes in the company currently assumes that such a manufacturing process is a given. As such, all other functions such as testing and quality control also design project plans that fit with the status quo manufacturing paradigm. As cost cutting ensues, most of the pressure falls on improving efficiency and that in turn makes the current processes more rigid.

To design and implement flexible processes, the company may have to build alternative R&D and manufacturing strategies from scratch. It may do so in parallel to the existing processes to avoid any loss of service. For example, the company can design a continuous process in which smaller amounts of materials are made continuously and tested, and the necessary information collected. In doing so, it will substantially reduce the lead time needed for large batch process and the information collected will allow it to make faster decisions to slow down, accelerate, or abandon the R&D program. To implement such a process, the company has to fundamentally transform the way it seeks, collects, and analyzes data. It also has to design equipment with switchability and modularity. In a fully modular continuous manufacturing process, the company will be able to configure the equipment to produce just the necessary amount of materials with low lead times. The equipment should also provide flexibility to shut down and start-up with relative ease and small warning windows.

To improve flexibility, the company has to also design more optionality into its processes. It has to analyze the need and effect of process optimization initiatives and consulting projects that may be going on. Most process optimization projects attempt to create lean and mean processes with little waste, low flex time, and high utilization. If done inappropriately, this can lead to rigid processes that may fail if the company faces uncertainties or a shock. The company has to weigh utilization and efficiency against flexibility and find the right combination. It is not to say that the company has to discontinue innovations in improving efficiency and productivity. It has to do so in the context of a flexible process. For example, the company can introduce process flexibility by designing contingent and parallel processes that it can switch to under certain conditions. By creating options to switch to alternative processes, the company can improve overall process flexibility while increasing the efficiency of status quo processes. Such a process design will help the company improve efficiency in normal times with low uncertainty but it can also switch to other designs if needed. Maintaining parallel and contingent options may be more expensive in the short run but if they aid the company's ability to weather shocks, it will have high economic value.

The content systems in the company are in disarray thanks to the many acquisitions it has undertaken. For a large company, this is an area that is most difficult to fix. The culture of a company develops slowly and it is a combination of many different components. Hiring and retention policies of the company affect its culture substantially. More importantly, to create a flexible and robust content system, the company has to clearly define what it stands for. The commonplace mission and vision statements that tend to read like fortune cookies do not necessarily reflect the company's belief system. Sending a few senior managers to off-site meetings to define the mission of the company is not the right approach. The company may want to start by asking all of its employees what their perspectives are. By taking information from all the employees, the company stands a better chance of reaching a well-defined belief system. This should lead to a set of principles that the company will always adhere to. For example, the company may have a principle of never bringing a defective product to market if the defect was known beforehand. This rule alone will help the company avoid the pitfalls of many of its peers. It is not quality controls, gates, regulations, rules, and operating manuals that differentiate good companies from bad. It is the presence of well-articulated principles. Since the company draws a blank in this area currently, it has to start with a blank sheet of paper involving all employees. The company has to consider the wishes of its investors as well. If the employees of the company define what the company stands for differently from the investors of the company, it may not be able to operate as it is. In this case, the company has to seriously consider dismantling itself and perhaps reorganizing into different companies.

How the company designs and maintains its legal and regulatory frameworks is also part of the content system. This also has to be driven by strong principles. For example, the company may define a principle of informing regulators of any mistakes as soon as it finds them. If it does not allow a qualitative evaluation of the severity of the mistake, this is a general principle. However, if the company has a set of rules and procedures that details when and how regulators will be informed when mistakes are made, it will have low clarity in its legal and regulatory systems.

The company has to develop clear principles for the development and maintenance of accounting practices and communications with investors. Even if the company's policy is to follow GAAP (generally accepted accounting principles) steadfastly, it does not mean that it has a strong content system in this area. GAAP, in spite of its name, is not a set of principles but rather a complex set of rules. Many companies who ran into trouble in the past followed GAAP. Since GAAP is a set of rules and careers are made in the legal profession finding creative ways to circumvent the rules, a culture of rules-based accounting may indicate weakness in this area. This is also related to how the company communicates with its investors. If the company has a principle of providing information related to its operations and strategy to its investors rather than numerical estimates of earnings per share (EPS) or gross margins, it will have a more robust accounting content. As most first year MBA students know, companies can easily dress up their financial statements to be fully compliant with GAAP. Supply chains and channels can make certain costs go away or show certain revenues magically. Such attempts by incompetent and ignorant corporate finance department will ultimately lead the company down a path to failure. The only way to fix the content systems in the company is to drive toward a set of clearly articulated principles created by the involvement of everybody in the company.

Internal strategy formulation and implementation have been a weak area for the company for decades. To introduce flexibility into its internal strategies, the company has to substantially change its ways of identifying problems and creating strategies. To identify internal problems, the company has to involve its employees. There is important information held by the employees. Some of this information is hoarded strategically but most are discarded as the company has no systematic process to surface them. Interviews and surveys based on templates are unlikely to bring this type of information to the fore. For example, the success of the company's R&D largely depends on its ability to identify successful prototypes early and fast track them. It also depends on its ability to abandon bad projects quickly. The selection and abandonment decisions on projects are typically practiced by committees. A committee of few people utilizing rigid analyses is unlikely to make optimal selection and abandonment decisions. The company has to create a mechanism by which all participants have the freedom to express how likely or unlikely a prototype is to be successful.

The stock market represents a mechanism of pricing an idea or a company. A large number of diverse participants making independent buying and selling decisions result in a reasonable estimate of what a company is worth. It does not matter if all participants in the market have complete information or if they make irrational choices. Similarly, an information market can be created within the company that allows a rating for the ideas it is pursuing. This will allow diverse employees in the company to express an opinion or feeling about the likelihood of success for various projects and prototypes. The wisdom of the crowd can be shown to be much more complete than a committee, however smart the committee members are. Such a process will be much more dynamic than the quarterly or annual committee meetings. The ability of the company to reflect changing and new information into decisions is an important component of strategy. No strategy is better than the one that considers all available information at any time than a static one created by experts, regardless of the ability of the experts.

Internal strategies also have to consider uncertainty explicitly. The company should abandon methodologies such as discounted cash flow (DCF) and decision tree analysis (DTA) that are too rigid. In both cases, deterministic cash flows are projected to the future and then discounted back to the present using arbitrary discount rates. Decision tree consultants have been active in large companies, able and willing to find a net present value (NPV) for anything as long as somebody can tell them the probability of regulatory and technical success (PRTS), precise costs, revenues, and timelines. The fact that some probabilities are binary (0 or 1) and that there are no precise estimates of costs, revenues, and timelines in R&D have not deterred them from applying these methodologies. They are also not bothered by the fact that an R&D program is a bundle of interacting options, giving the company flexibility to alter cash flows in the future by abandoning, slowing down, accelerating, or switching programs.

The company has to apply generalized methodologies that allow the incorporation of uncertainty and decision flexibility. The company should focus on managing risk holistically rather than creating arbitrarily rigid NPVs and pretending to make decisions that way. Internal or real risk emanate from products, technologies, people, processes, and prototypes. There are many interrelated components in the company's portfolio and hence portfolio risk is very different from the risk at the component level. To make matters more complicated, the real risk also interacts with financial risk that emanates from its capital structure and the overall market. To create robust internal strategies, the company has to create a holistic risk management system that cuts across the entire company. In shaping its portfolio, the company has to consider both downside risk and upside potential. For example, if the company's portfolio has high downside risk, it can lead to catastrophic failure in a shock. This can happen if the company has concentration in certain areas or if the portfolio is driven by one or two large products, the failure of which is synonymous with the failure of the company. Even though traditional financial metrics may give the company high ratings, the accumulated risk and rigidity may push the company to failure in a shock.

In improving flexibility in its boundary strategies, the company has to robustly link its internal strategies with that of partners, collaborators, and customers. The licensing and sourcing contracts have to consider both its own and the partner's internal risk and flexibility. Features have to be designed so that the partnerships allow the company to effectively manage uncertainty and dampen a shock. Partnerships and collaborations have to consider location, type, timing, and other components that may allow the company to diversify risk more optimally. It has to also challenge existing practices that may have been designed under a previous regulatory regime and may not be relevant currently. The company should redesign its regulatory strategy and interactions so as to take maximum advantage of contemporary policies and information.

In thinking about external strategies, the company may have to redesign its marketing and sales infrastructure from scratch. Since it was designed based on a set of constant assumptions from a decade ago, it may not be relevant anymore. Given the uncertainty in R&D productivity and regulatory posture, the company should reduce the fixed costs in its current marketing and sales organization. By creating flexible contracts with external marketing organizations, it can better manage uncertainty and reduce overall risk. It should challenge the need for a large internal sales organization that increases fixed costs and operating leverage. Because of uncertainty in R&D, products reach market in waves and the company has gone through long periods of drought. Having an internal sales force in this case will reduce flexibility. The company should consider developing a more flexible sales force, part of which may be assembled through partners or even competitors.

Figure 8.3 contains a project plan for the transformation of PureTech into a more flexible company. PureTech can introduce significant flexibility into its structure, systems, and strategy in the next five years.

Years	1	2	3	4	5
De-layer the company by eliminating middle management layers	■	■			
De-layer the company by combining bottom layers and eliminating traditional titles	■				
Reduce age and gender concentration by rotating employees through layers		■	■		
Reduce compensation leverage by lowering fixed compensation in top layers	■				
Reduce compensation leverage by introducing restricted stock based compensation in bottom layers	■				
Redesign hiring and retention policies that involve a larger number of employees	■				
Create smaller, autonomous and complete (include all functions) units with SOUL characteristics	■	■	■		
Relax specific education/experience requirements in hiring and seek complementary skills within units	■				
Encourage movement and participation of employees across units	■	■	■		
Institute democratic election process for selecting executives and managers of the company		■	■	■	
Define a decision hierarchy for the entire company	■				
Create an information structure with context that supports the decision hierarchy		■	■		
Unify information structures across departments and units			■	■	
Find a technology partner to manage all information technology and outsource it		■	■		
Create non-proprietary hosting and management of information using cloud based computing				■	
Integrate technology systems in the creation of a smart company and factory				■	■
Redesign supplier contracts to introduce more flexibility with tiered pricing and insurance		■	■		
Introduce modularity and switchability in physical infrastructure with new equipment and facilities			■	■	■
Eliminate static and yearly budgeting process and incorporate dynamic resource allocation	■	■			
Utilize cash abroad for infrastructure redesign and collaborations			■	■	■
Create a set of operating principles and cultural norms for the company, involving all employees	■	■			
Eliminate EPS forecasting and provide investors operating metrics with transparency	■	■			
Define principles (not rules) for the communication of information to regulators and investors	■	■			
Create an information market with employee participation for the selection and design of projects		■	■	■	■
Eliminate analyses that do not consider uncertainty and contingent flexibility		■	■		
Integrate real and financial assets in a company-wide portfolio		■	■	■	■
Create or redesign partnerships and collaborations considering location, type, and timing		■	■	■	
Eliminate fixed costs in sales and marketing infrastructure			■	■	■
Eliminate corporate offices and create regionally autonomous units			■	■	■

FIGURE 8.3
Plan of action.

9

Creating a Flexible Company

A flexible company is one that is able to manage and take advantage of the unavoidable, and not to be avoided, uncertainty through flexibility. Flexibility is a holistic notion that permeates every aspect of a company. An entrepreneur who is creating a new company has a great opportunity to design a flexible organization.

The first component to focus on is the human resource structure. The ideal structure, as we have already seen, is a SOUL. However, in many cases, the conventional templates are followed. The capital provider, typically a venture capitalist (VC), may insist that the entrepreneur has the "team" in place. The team may include the CEO, COO, CFO, CSO, and other Cs. These labels and functions seem to provide a level of comfort for the VC that somebody is in charge. However, in the process, they may have architected rigidity into the company, and this also may ultimately result in the failure of the company. If the first act of establishing a company is creating layers and specialization, it is unlikely that the company will succeed. What the company needs is a group of people to come together who believe in a concept and share a passion to build the company. The creation of titles and the recruitment of people based on "previous experience" will result in a rigid human resource structure for the young company. Many believe that there is a special breed of people willing and able to make companies successful. Apparently, they approach the creation and development of companies mechanistically, and their success has nothing to do with whether or not they believe in the company's ideas. Given the rather meager returns posted by the venture capital industry this decade, it is unclear if it works this way.

Because the capital provider is likely to impose a rigid human resource structure, the entrepreneur has to be selective in raising capital. Even if the capital provider shows flexibility in the creation and organization of the nascent company, the entrepreneur could fall into the same trap. Most entrepreneurs believe that they are special and they should be running the company. In information-driven industries such as life sciences and high technology, the founders generally have a technical background and the company is based on technical ideas. However, being a great technologist does not automatically qualify somebody to run a company. The ego of the entrepreneur drives him toward creating a layered structure in which the person posts himself as the CEO. The end effect of this is the same as the structure imposed by the capital provider. Entrepreneurs should resist the urge to appoint themselves as pyramidal leaders and create pyramidal structures if they really want to succeed. They also should avoid capital from those sources that require a pyramidal structure as a necessary condition for the provision of capital.

Capital providers also believe that companies they invest in should create an infrastructure—an office, a lab, and other paraphernalia. The idea is that it is better to manage people if they show up at some central location. The seeds of infrastructure rigidity are sowed early and the company will progress down a path of bigger owned infrastructure and correspondingly lower infrastructure flexibility over time. The availability of people at a central location allows the company to gather them into meeting places and create procedural management requirements that have no relevance to what the company is trying to accomplish. From the entrepreneur's perspective, it is important to design the company based on its objectives. Entrepreneurs should resist all temptations to establish artifacts based on established templates. We have seen that a networked organization is likely much more flexible than an owned and established infrastructure at a centralized location. If the company were to establish a physical infrastructure, it should keep modularity and switching ability as important design components. It should invest in these attributes and should not attempt to economize. Since the company has to manage through uncertainty, it is imperative that it has sufficient flexibility to do so.

Many entrepreneurs tend to seek the bare minimum amount of capital from investors. This is because the investors demand more of the company if they were to invest more. This tension—the entrepreneur trying to minimize what is being given away and the investor trying to grab as much as possible—almost always leads to a suboptimal capitalization of the company. Once the company is capitalized at less than what it needs to be successful, it will try to economize in many aspects—R&D design, collaborations, brand building, hiring—essentially guaranteeing failure. Both the entrepreneur's and the investor's attempts at maximizing their own shares of the company end up in one of them holding more than the other of an essentially valueless asset—a failed company. Another common aspect of the term sheet is that the investors fill it up with many hard to value features such as anti-dilution protection. In doing so, they siphon out more of the apparent "value" of the company to their pockets. The entrepreneur may not realize this at first but it becomes clearer later and this realization destroys all motivation for performance. Once again, short-sighted, tactical, and ignorant actions of capital providers result in them holding a higher percentage of something that is worth nothing (failed company) rather than a smaller percentage of something with some value (successful company).

There are also many games played between the original investors and the new investors in companies. Both try to grab a higher percentage of the company to maximize their take. Board meetings of these companies tend to be contentious with different classes of investors vying for a larger share. In the process, the company and its objectives may be delegated to secondary status with disastrous end results. Entrepreneurs wanting to build successful companies should avoid such a process at any cost. They should seek the right level of capital needed from an investor who is interested in the objectives of the company. Rather than focusing on maximizing the percentage holding of the company, the entrepreneur should put all efforts into finding the right investor, willing and able to provide the right level of capital to make the company successful.

How the company collects and uses data early on in its inception has implications for how the information structure will develop later. For example, many investors are driven by "numbers," and they may force the company to collect data in many aspects even though it has no relevance for the decisions it will make. Board meetings may be filled with people throwing statistics at each other as if that advances the fundamental premise of the company. Cheap computer hardware and even cheaper software for data collection, dissection, and presentation create huge traps for the company to fall into as it establishes itself. By understanding the decisions that are important and the data that will support the decisions in a top-down decision hierarchy will help the company avoid big data collection and storage efforts and associated information rigidity.

The company should also focus on designing systems that integrate its core operations with business. A blind implementation of information technology systems that does not integrate with its R&D and manufacturing (if they exist) will initiate a spiral down in information system inflexibility. The systems architecture should be minimal and should focus on the core aspects of the company. Start-up companies should also consider using systems that are nonproprietary and able to advance and change with the time. By housing data in an information cloud outside the boundaries of the company and utilizing established systems provided by technology specialists will help companies avoid creating their own systems. This will also provide them with systems that are state of the art and flexible. Such systems will be able to incorporate new ideas developed elsewhere and provide new companies a path to the future as their own needs change.

If a new company is established by people who worked in large corporations, the tendency will be to create a process-centric company. As we have seen, efficient processes in the absence of flexibility can be a disaster. For a new company, flexible processes are much more important as it may be innovating at a much higher rate in areas that are critical to the success of the company. In designing processes, a new company should also reject established notions of complexity—such as multiple locations, products, and other aspects. For example, companies may consider processes that encompass multiple countries to be too complex and difficult to manage. This efficiency focus will be exactly the wrong medicine for a new company. It should seek out opportunities to connect to other networks and keep all options open as long as possible. The company should create contingent and parallel processes in a redundant way to have a higher level of optionality, even if it is a more expensive way to manage. Since a new company faces a higher level of uncertainty, it has to design a much higher level of flexibility than an established company to avoid catastrophic failure.

Establishing a strong culture and ethical standards of the company upfront is extremely important. There should be a set of principles the founders of the company articulate clearly. They should be understood by both the investors and the participants in the company. This will allow the company to perpetuate them consistently as it grows. Founders of the company should resist creating elaborate rules, procedures, and operating manuals as they can constrain creativity and innovation. The founders also should be willing to take swift actions against those who are not a good match with the company's principles and culture. Compromising on established principles for any reason will make the company weaker and set a bad precedent. Over time, this will destroy the content systems in the company and force it to create more and more rules and procedures in an attempt to curb risk.

In creating strategies for young companies, the founders should be keenly aware of pervasive uncertainty and the power of flexibility. They should avoid the temptations of managing the company using rigid financial forecasts. For example, investors in the company may require pro-forma financials. The income statement and cash flow statements require projections of the company's costs and revenues. It also requires specific information on timelines, products, margins, and other aspects. Entrepreneurs may do such analyses and create financials either using an optimistic scenario or an average scenario. They may also keep a pessimistic scenario in their back pocket, just in case they are challenged on their assumptions. Creating financials with static projections and then multiplying such analysis manyfold—to incorporate uncertainty—is a bad way of managing a company. Instead, managers of the company should get comfortable with using uncertain inputs in their models. More importantly, planning should incorporate flexibility such as contingent plans and the ability to switch between alternatives. Developing this way of thinking will help the company create a culture of more robust strategy setting through the management of risk.

In summary, entrepreneurs and founders of new companies should seek the right level of capital (not less) from the right investors (who participate because of their interest in the company's objectives); design a flat organization with the right incentives for performance (avoid the use of traditional templates and executive titles); settle for a minimal physical infrastructure with future expandability, switchability, and modularity; seek and collect just enough data to support the important decisions in the company; favor flexible and redundant processes against efficient and singular processes; develop a strong cultural and ethical content by basing the company on a set of clearly articulated principles; and consider uncertainty and flexibility in all strategies pursued by the company. If so, they will initiate a flexible company with a higher probability of success. They will also be able to retain the flexibility as the company grows by avoiding the common pitfalls in structure, systems, and strategies.

10

Flexible Countries and the World

Macroeconomic and social policies have an adverse impact on the overall flexibility in the structure, system, and strategy of a country. Countries face uncertainty, very similar to the companies within it. This may be related to its demographics, resources, trade, currency, and other aspects as well as its interactions with other countries in the world. Policies that reduce flexibility in any of these components may reduce overall value, utility, and quality of life for the citizens of the country.

A country with a higher diversity in its people in terms of age, race, religion, and other aspects will be more flexible than others. For example, if the country has a higher percentage of retired people compared to the cohort that is currently in the workplace, it will have lower flexibility due to predictable and high levels of fixed cost. Countries with low age flexibility can improve the situation by relaxing immigration rules for students. Policies that attract older workers back into the workplace are also advantageous in this regard. Traditional ideas of retirement and other age-related planning result in a loss of flexibility for the individual, the company, and the country. A fixed retirement age is one such policy that has a negative effect on flexibility. This can be mandated by the government or accepted by the populace at large as tradition. In either case, most individuals "plan" to retire at a specific age, and this may lead to suboptimal decisions at all levels. With innovations in health care, past expectations of life span and retirement age may have to be rethought. An individual can continue to contribute to society much further into her life. Age-related hiring and retention policies in pyramidal companies are partly responsible for this outcome. A constant retirement age in current companies also means that the pyramidal leader will have to amass enough wealth to "comfortably retire." Policies that discourage such practices may be one way to increase age flexibility to the whole country and the economy.

Race is a controversial subject. The United States provides a great example of race-related diversity. From its inception, the country attracted many different races and people, leading to the creation of a thoroughly mixed and diverse group. Countries with restrictive immigration policies, or countries that did not have sufficient incentives to induce immigration, lagged. Uniformity in thinking and culture is generally a less attractive attribute in the context of flexibility. Immigration has been fast and furious for most of the last 50,000 years, since humans arrived on Earth. For the past few thousand years, the establishment of clans and then countries has created immigration barriers across the world either through restrictive policies or through natural barriers such as language and culture.

Another controversial subject, religion, has an equally important influence on a country's flexibility. A country that is organized under a single religion will be less flexible compared to the one that is secular and allows many religions to thrive. Religions generally have a codified set of rules and influence uniformity in ideas and thought processes. There is also a risk of creating self-reinforcing feedback processes that continue to build a set way of thinking. In some countries, such feedbacks surface as religious extremism—an inability to appreciate differing perspectives and ideas. Exposure to other countries/cultures and education can provide some relief in this dimension.

A country that is democratic will have higher flexibility compared to other forms of governance. This is because democracy allows faster incorporation of new information into the governance structure. Countries with autocratic regimes resist the identification and incorporation of new information. In fact, autocratic regimes have great incentives to suppress all new information. Democracy thus is the optimal way to reflect and adapt to new information and is likely the best system under uncertainty. Direct democracy, where all members of the society take part in decision making, is the best form of democracy. This is because the participation of all members guarantees that the decisions are made with complete information. This form of democracy was practiced in the fifth century BC in Athena. Although this was not a pure direct democracy, as only males were allowed to take part, it was the first attempt at improving decision quality in governance. In modern representative democracies, there is information loss due to lack of choices. Because traditional electoral processes are elaborate, time consuming and expensive, they tend to attract people with certain defined characteristics. This means that in most democracies, the elected representatives seldom match the overall demographics of the country. Elected representatives may show high concentration in gender, age, and class. This unavoidable concentration, due to the existing incentives in the system, creates a rigid body, unable and unwilling to represent the insights and wishes of a larger and more flexible population. In the modern world with the availability of Internet and related technologies, countries may want to return to a form of direct democracy again.

A country with a higher percentage of its GDP derived from international trade will have a higher level of flexibility. This is because international trade allows the country to diversify its risk in a variety of products, services, and locations. It also allows the country to be exposed to a larger set of new ideas and this enhances the rate of innovation. High levels of trade combined with a free-floating currency allow the country to dampen shocks and thus manage uncertainty better. A pegged currency or a fixed currency exchange rate regime removes the automatic shock absorbers in the country's interactions outside its boundaries. For example, a country with a large current account deficit will face a depreciating currency, automatically making its own products and services more attractive abroad, enhancing exports, and in turn reducing its current account surplus. These automatic adjustments will be continuous and incremental, always keeping the country in equilibrium. Any constraint on such adjustments will create an artificial environment that will continue to accumulate stress. In such an environment, inefficient industries may thrive aided by the artificial currency rates. Just like a geographical fault accumulates stress and catastrophically releases it in an earthquake, fixed or pegged currency regimes will eventually fail with a sudden devaluation. The companies and industries in the protected country will be less able to deal with such a shock as they are constructed under a different set of assumptions. This has happened to many developing countries in the past, where ignorant bureaucrats ran artificial currency regimes on the belief that they can strategically engineer optimal growth rate for them.

A country, with a clearly defined set of property rights, operating in a free market system will be more flexible than the one where no markets exist and production and wealth acquisition rates are set by a central entity (as in centralized planning). This is because a free market system will automatically divert resources to higher return activities as new information becomes available. Centralized planning, run by a small group of people, will be slow to both recognize new information and act on it. The suboptimal decisions made by the leaders in centralized planning regimes resulted in overinvestment in failing and inefficient industries and underinvestment in those areas where competitive advantages existed. The outcome of bad management is always failure in these countries just like today's companies run by pyramidal leaders.

A country with a larger percentage of investments in sectors with expected payoffs in the future has a higher level of flexibility than one that is focused on today's demand and supply of established products. In the industrial context, this means that countries that produce higher value-added products and services likely have higher flexibility. This is because companies at the end of a supply chain have higher flexibility to switch between providers of raw material inputs and energy. Since raw materials and traditional energy-carrying resources such as crude oil and coal show stable and rigid characteristics in production, logistics, and use, they are unable to change as demand patterns change. The producer of the end product (that uses the raw materials and energy) can more easily switch to another product as more information becomes available. Thus, economies driven by mining and crude oil production show low flexibility and are less able to cope with uncertainty. This is also true in the service sector where commodity activities show low flexibility as opposed to value-added services. For example, a country that specializes in "call centers" that is a commodity activity will be less equipped to deal with uncertainty compared to the one that uses such "call centers" as an option in its service process. The end user also retains location and timing flexibility, with the ability to switch the commodity activity from one place to another as the situation demands, perhaps driven by demand/supply patterns or currency movements.

A country that is active in many different industries and services as opposed to a few specialized ones has a higher level of flexibility. A diverse set of products and services allows the country to switch the focus and concentration among them based on evolving conditions. For example, a country that makes both cars and airplanes has a higher level of flexibility compared to the one that makes just one or the other. If the country makes both, it can shift employment from one to the other if the demand for one declines and the other is steady or growing. Diverse industries also provide a country with cross-industry learning and the ability to innovate from it. For example, a country with both pharmaceuticals and chemicals industries may be able to apply learning from a cost competitive industry such as chemicals to the manufacturing processes in pharmaceuticals that has a focus on intellectual property.

A country with a large number of small companies as opposed to a small number of large companies is more flexible. A large number of small companies allow the country to replicate a flexible system, akin to the uni-layer human structure inside companies. Each small company has a better alignment of incentives as well as a set of competencies that may be different from others. This provides a higher level of diversity and thus flexibility for the country. For example, a country where a single employer (say, mobile telephone maker) represents a large percentage of workers is less able to manage uncertainty compared to another that has a number of small companies active in many different industries. If the demand for mobile telephones falls, perhaps due to the invention of a new type of communication device, countries with high industry concentration in traditional telephony will be affected more than others. If the industry concentration is accompanied by a single monopoly or a monopolistic structure, the country will be in a worse position as survival and success depend solely on a few companies. If such companies are set up in a traditional pyramidal fashion, the country's future prospects will be largely driven by the ability and decisions taken by a few people at the top of the pyramid. As we know from past experience, such processes are unlikely to be optimal in the face of uncertainty.

A country that has a well-defined legal system and consistent implementation of all laws across the population without discrimination will have high flexibility. This is because such a country will eliminate fraud at all levels—government, institutions, and individuals—and provide the population with a high level of confidence in the structure, system, and strategy of the country. Fraud is the ultimate flexibility reducer both when it is perpetrated and when it is discovered. If fraud is perpetrated, it implies that there are breaks in the system that allows somebody or some group to act in violation of the law. Such actions would certainly have created nonoptimal outcome for most participants. When the fraud is discovered, it has a negative effect on the confidence of all participants. For example, if the fraud is discovered in the government, people may begin to question and second-guess all actions of the government. If the fraud is discovered in the financial system, it may lead to lower investments, hoarding of zero return cash and associated decline in flexibility and innovation rate. Thus, design and consistent implementation of a well-articulated and clear legal system is necessary to ensure against sudden loss of flexibility due to the perpetration and discovery of fraud.

If there are implicit or explicit policies that aid a certain type and size of companies or individuals in the economy, that also can be considered fraud at a grander scale. For example, if a company or financial institution enjoys exceptions and favorable treatment from the regulators because of its size, importance, or proximity, this will lead to government-induced market failures. A consistent implementation of regulations will reduce uncertainty and provide all companies with the right incentives to innovate. Similarly, if implicit or explicit government guarantees are given to certain companies, that can lead to bad decisions, decreasing flexibility and increasing risk.

A country that has a passive monetary policy that targets a clearly articulated money supply by automatic mechanisms to keep it within bounds will have a higher level of flexibility than one in which the monetary authorities constantly try to tweak interest rates or other attributes. This is because automatic adjustment mechanisms are much faster to react and are not prone to the whims of grand visions of monetary authorities. By reacting faster to new information, the system will never move far into extremes. If decisions are made by a few, however smart they are, the system will be less flexible compared to the one where decisions are influenced by the actions of a large number of people. If policy makers keep interest rates too low for too long, the country will be awash with money and associated inflation, forcing policy makers to take further actions, removing flexibility. If policy makers keep interest rates too high for too long, they can drive the country into a deflationary spiral, forcing unnatural actions by companies and policy makers, thus removing flexibility. A passive and well articulated monetary policy with automated correction mechanisms allow small corrections as new information becomes available and will always keep the country and the companies within it in a well understood regime.

A current account and a budget that is neutral and fairly stable also add flexibility as a country can better weather shocks in either direction. Just like companies, low debt in the capital structure also allows higher flexibility for a country in managing GDP growth. Too high current account/budget deficit or surplus indicates a system that is not in balance and thus less flexible. As the readers know, current account deficit naturally means a capital account surplus. So, the ideal situation for a country is to have neither current account surplus nor deficit. This can be accomplished by running a balanced budgeting process.

A consistent and neutral fiscal policy adds more flexibility to the economy. A tax system that provides incentives for investment and innovation will add to the flexibility as it will increase the country's ability to create new products and services to cope with uncertainty. Tax policies that target income, in general, will reduce incentives for innovation and thus reduce flexibility. Countries with high income taxes may reach a stage of no innovation and low incremental investment and low flexibility. Such countries will be less able to cope with external shocks. For example, a country that provides its citizens with high levels of government services through high income taxes may work well in a stable macroeconomic regime. The population of such a country will tend toward uniformity in all aspects and this will reduce overall flexibility. Since such policies also result in less investments and innovation, the country will be unable to cope with a shock. In periods of relative stability, such a country may appear to have a high quality of life compared to other countries with low taxes and low government involvement. However, when a shock, such as a loss of demand for the country's major export, happens, the country will be unable to cope with it, resulting in significant pain during prolonged readjustment.

A tax system that is neutral (neither progressive nor regressive) may be optimal for increasing overall flexibility. A simple tax system with no income taxes and a neutral consumption tax will help induce savings and investments, leading to a higher level of innovation. A country that saves and invests has a higher level of flexibility due to capital formation and the impact on innovation compared to others, who may just save or just consume. Taxing income provides a disincentive for the creation of income and thus will dampen investment. Taxing consumption, on the other hand, will increase savings, which in turn result in capital formation and sustainable future investment. Many object to a consumption tax because of its effect on the progressivity in taxes. This, however, is an implementation issue and can be solved in many different ways. For example, a consumption tax exemption for basic food, education, clothing, housing, and health care can make it progressive. Policy makers should not be deterred from selecting the best tax policies due to political considerations.

A country with lower tariffs, subsidies, and other government-introduced constraints will be more flexible. Many countries provide blanket subsidies for industries they believe to be strategically important. If such subsidies are designed to encourage the prolongation or creation of inefficient industries, the country will lose flexibility in the long run. Policy makers can, however, design creative ways to provide subsidies to induce optimal investments.

The culture and makeup of a country is an important determinant of the level of flexibility. For example, established cultures with a long history tend to have fairly rigid customs and hierarchy. In many Eastern cultures, age is considered to be the primary determinant of wisdom and questioning elders is taboo. Such a society will have low flexibility and likely lower innovation rates. However, it is quite possible to construct a society that takes advantage of both flexibility and traditions. For example, traditions provide a framework for people to come together and may provide better conditions for the establishment of well-connected networks. If the hierarchical aspects of the society can be overcome by taking advantage of the availability of large networks, it may provide opportunities for the creation of a more stable but still a flexible society. However, countries with longstanding traditions try to maintain status quo because of the fear of losing their culture. In this process, they may reject new developments and the adoption of more flexible approaches to organizing.

Ever since the first human stood up in the African savannah, the story of humans has been one with many twists and turns. Humans ventured out of Africa, either because of the changing climate made food and water rare or they were looking for new adventures out of curiosity. In their first migratory paths into the Middle East and Asia, they would have encountered uncertainty in climate, food availability, and dangers from animals. However, they survived and thrived, perhaps indicating a level of flexibility that was inherent in the way they gathered information, organized, and lived. The larger brains have given humans a higher ability to process new information faster, giving them an edge in analyzing nonlinear and interacting data. The innate ability to store and retrieve large amounts of information in real time further added to the ability of humans to make complex decisions faster. Through the unique innovation of language, humans have also been able to transmit information from generation to generation, setting them apart from most other animals. All of these uniquely equipped humans to innovate and thus become more adept at managing uncertainty. It is this flexibility that allowed them to escape the harsh and sudden climatic changes and spread across the globe.

It may also have been pure luck. Earlier in human history, there are indications that *Homo sapiens* survived a near extinction, a spectacular bottleneck in human history, with a few thousand surviving in the entire world. Such an event would have caused uniformity in the genetic makeup of modern humans. Most studies indicate that humans in the world today show very low genetic diversity. Such a close escape from extinction shows that humans indeed are vulnerable to sudden changes in the environment or disease. This has implications for the human race as they design future systems for their own survival and possible journey out of Earth. Thus far, the focus has been mainly on the design and powering of transportation mechanisms for such an eventuality. However, given the biological rigidity in the human race, we have to also pay attention to how to nurture whatever diversity is present. It is not just about the transportation mechanism but also what is being transported. Policies that drain biological diversity will drive the human race to uniformity, increasing the risk of extinction or reducing the ability to escape from the Earth in a catastrophe.

Ever since humans arrived on Earth, they have been organizing into clans and groups for safety and scale in hunting and gathering. Over time, the small groups grew into large societies and civilizations. Most of us look back to the ancient civilizations in awe—at the sheer scale of what they have accomplished. But many of the great civilizations failed in what appears to be a familiar cycle—small groups growing into larger ones, ultimately ending up in gigantic civilizations, and finally perishing when unexpected events—such as sudden climatic changes—occur. Although the cause and effect is difficult to determine, one possibility is that a larger system is less flexible and less able to cope with sudden changes. Most ancient civilizations were also prescriptive, driven by a few people at the top governing through a strict hierarchy. There were no attempts at maximizing societal utility and most resources were diverted into meaningless and destructive activities such as the construction of monuments and waging of wars. Although such systems are able to perpetuate themselves in periods of low uncertainty, they are not equipped to deal with shocks. In the last 2,000 years, humanity has been able to organize into large blocks and build civilizations at a grander scale. Humans have successfully inhabited all corners of the world and multiplied into over six billion individuals. In the process, we have divided ourselves in many dimensions, countries, religions, and classes based on color, gender, age, and wealth. Each of these blocks, with defined rules and regulations, represents a rigid attribute for the modern society. Thus, we have been able to take the attributes of the failed ancient civilizations and multiply it manyfold, stuffing ourselves in a sort of straitjacket, losing all flexibility.

Since innovation is the primary provider of flexibility in modern systems, structures that allow maximum innovation are likely the best. Innovation is driven by information and so unencumbered transmission of information system-wide is a necessary condition to maximize it. Thus, any structure that compartmentalizes and reduces information flow will be suboptimal. This is not to say that a country with a unique culture or a religion is bad. It is the lack of information flow and the ability to share and innovate to solve common problems that causes rigidity.

To introduce flexibility into the modern society, we have to break the shackles of compartmentalization that has been the norm for the past few thousand years. Given the initial conditions—segmentations into noninteracting countries, religions, and classes—this is not easy.

To thaw the frozen world, one has to attack all of these problems simultaneously. For example, those lacking basic necessities should be given capital and incentives to grow out of that predicament. The instinctual reaction of those who see the problem and want to solve it has been to shower food and money on those who do not have it. This policy has actually taken people in these countries away from solving the fundamental issue—a lack of capital and incentives to solve the problem themselves. Recently, some forward-thinking entrepreneurs have attempted to move in this direction by providing micro-loans to people in underdeveloped countries to jump start innovation and problem solving. This is the right approach and can finally move underdeveloped countries to a point of self-sufficiency, multiplying the level of creativity and human potential available for the world.

The second problem—ignorance—is primarily due to lack of education. Although, education is not a guaranteed cure of ignorance, it is the first good step. Education does not mean traditional tracks such as college and graduate school. The primary part of education that will alleviate ignorance is a holistic exposure to the world—either through travel or through independent media. Part of the knowledge that everybody should gain is the understanding that the six billion people who currently inhabit the world differ very little in their fundamental makeup. Obvious surface features such as color, height, and weight are not relevant, but they figure prominently in the thoughts and actions of most people today. With the advent of the Internet and wireless technologies, the world has been tied together more closely. This has allowed the younger generations to not be constrained by proximity and familiarity in the formation of groups and partners. Discussing and sharing new ideas and collaborating to innovate do not have to be with people who are close by anymore. The playground has literally expanded to the whole world, and this one aspect is introducing flexibility into the way people organize, think, and act.

As innovation is the most efficient way to introduce flexibility, leaders and policy makers have to create an environment that encourages it. To introduce flexibility in the country dimension, one does not have to eliminate countries but rather create systems that allow a high level of interaction among them. Free trade is an essential aspect of connecting countries together. Free movement of people between countries is the necessary next step. Countries have to ultimately drive membership by subscription and not by birth or compulsion. In such a system, the location of an individual will be driven by the desire to be there because of the climate, food, culture, and other aspects. If so, the inflexibility emanating from the country dimension can be virtually eliminated. Large multinational companies behave more like countries today. Although they provide a valuable service by cutting through the conventional boundaries of countries, they may create rigidity in other ways by raising barriers or influencing protectionist policies.

The rigidity coming from the religious dimension is more difficult to solve in the current state. Since religion is based on faith, it is vulnerable to be manipulated by a few in positions of power. Since most religions today are organized and follow a defined code, they tend to change very slowly. This inflexibility drives most religions to be out of sync with the current environment. To introduce flexibility in the religious dimension, the leaders have to embrace the present and make appropriate changes to the initial ideas. A unification of the fundamental ideas of the various religions will be the next step. As many of these differ mostly in the interpretations and procedures and not in the content, it is rational to unify them into a single set of ideas.

We can learn much from the success and failure of previous civilizations. One attribute that sets humans apart from other animals is our ability to analyze past data and learn from our mistakes. We have to progressively unwind the complex and rigid system, first allowing a higher level of flexibility in the class dimension through better education and better economic policies for developing countries. The country dimension can be made more flexible by free trade and free exchange of ideas, first through larger trade blocks and then a system that encompasses the whole world. Finally, the religious dimension has to be made more flexible by unifying varying interpretations of the same ideas and systematically removing suspicion and hatred based on lack of information. If we are unable to introduce more flexibility in all these dimensions, we are at risk of failure, just as previous civilizations that had similar issues.

We have established that flexibility is fundamentally important for individuals, organizations, countries, and the world. As we have seen in the examples given, flexibility at the individual level can be improved by acquiring a larger portfolio of skills, continuous education, creating a broader perspective by understanding the world, having access to financial resources, having good health, being able to move to various locations, speaking different languages, and in general keeping an open mind to new ideas. Organizations that attract individuals with flexibility satisfy the necessary condition for flexibility within it. In addition, an organization has to impart flexibility in its human and information structure, real and financial infrastructure, technology and processes systems, cultural and ethical content, and the strategies it follows. Having flexible organizations—both public and private—improves the flexibility of the country they are part of. In addition, to be flexible, a country has to improve its diversity in race, religion, and industry; reduce concentration within industries; allow freedom in trade, markets, and capital creation; and follow passive fiscal and monetary policies so that the system can adjust to new information quickly. Flexible countries can form trade blocks and continental alliances to take advantage of proximity creating large and flexible unions of countries and cultures. Ultimately if the world can remove the barriers of countries, religions, and classes that exist today that divide the population into rigid blocks, it can reach an optimal state of flexibility. This will help us reach a more productive state and make the next leap in the evolution of human thought and spirit.

Bibliography

Ash, M. G. (1992). Cultural contexts and scientific change in psychology: Kurt Lewin in Iowa. *American Psychologist, 47*(2), 198–207.

Baron, R. A., & Greenberg, J. (2008). *Behavior in organizations* (9th ed.). Upper Saddle River, NJ: Pearson Education.

Becker, L. C. & Becker, C. B. (Eds.) (2002). *Encyclopedia of Ethics*. Second edition in three volumes. New York: Routledge.

Buer, M. C. (1926). *Health, wealth and population in the early days of the Industrial Revolution*. London: George Routledge & Sons.

Butchvarov, P. (1989). *Skepticism in ethics*. Bloomington, IN: Indiana University Press.

Castells, M. (1996). *The rise of the networked society: The information age*. Cambridge, MA: Blackwell Publishers.

Chandler, A. (1962). *Strategy and structure: Chapters in the history of industrial enterprise*. New York: Doubleday.

Coase, R. (1937, November). The nature of the firm. *Economica, 4*(16), 386–405.

Corner, P., Kinicki, A., & Keats, B. (1994). Integrating organizational and individual information processing perspectives on choice. *Organizational Science, 3*.

Drucker, P. (1954). *The practice of management*. New York: Harper and Row.

Drucker, P. (1969). *The age of discontinuity*. London: Heinemann.

Dunnette, M. D. (Ed.). (1976). *Handbook of industrial and organizational psychology*. Chicago: Rand McNally.

Eatwell, J., Milgate, M., & Newman, P. (Eds.). (1987). *The new Palgrave: A dictionary of economics*. London: Macmillan and Stockton.

Hatch, M. J. (2006). *Organization theory: Modern, symbolic, and postmodern perspectives* (2nd ed.). New York: Oxford University Press.

Johnson, R. A. (1976). *Management, systems, and society: An introduction*. Pacific Palisades, CA: Goodyear.

Keynes, J. M. (2007). *The general theory of employment, interest and money*. Basingstoke, Hampshire, UK: Palgrave Macmillan. (Original work published in 1936)

Koppes, L. L. (Ed.). (2007). *Historical perspectives in industrial and organizational psychology*. Mahwah, NJ: Lawrence Erlbaum.

Kotter, J. (1982). *The general manager*. New York: Free Press.

Lee, E. (1996). Globalization and employment. *International Labour Review, 135*(5), 485–498.

Liebeskind, J. P. (1996, Winter). Knowledge, strategy, and the theory of the firm. *Strategic Management Journal, 17*.

Mintzberg, H. (1973). *The nature of managerial work*. New York: Harper and Row.

Moore, M. H. (1995). *Creating public value: Strategic management in government*. Cambridge, MA: Harvard University Press.

Morville, P., & Rosenfeld, L. (2006). *Information architecture for the World Wide Web*. Sebastopol, CA: O'Reilly.

Ohmae, K. (1982). *The mind of the strategist*. New York: McGraw Hill.

Reed, M. I. (1985). *Redirections in organizational analysis*. London: Tavistock Publications.

Robbins, S. P. (2004). *Organizational behavior—Concepts, controversies, applications* (4th ed.). Upper Saddle River, NJ: Prentice Hall.

Rogelberg, S., G. (Ed.). (2002). *Handbook of research methods in industrial and organizational psychology.* Malden, MA: Blackwell.

Sanders, Ross, & Coleman. (1999, July). The process map. *Quality Engineering, 11*(4), 555–561.

Schuck, G. (1985, Autumn). Intelligent workers: A new pedagogy for the high tech workplace. *Organizational Dynamics.*

Selznick, P. (1957). *Leadership in administration: A sociological interpretation.* University of California Press.

Simon, H. A. (1997). *Administrative behavior: A study of decision-making processes in administrative organizations* (4th ed.). New York: Free Press.

Stone, R. (2005). *Human resource management* (5th ed., pp. 412–414). Queensland, Australia: John Wiley & Sons.

Sullivan, A., & Sheffrin, S. M. (2003). *Economics: Principles in action.* Upper Saddle River, NJ: Pearson Prentice Hall.

Sweet, F. H. (1964). *Strategic planning: A conceptual study.* Austin: Bureau of Business Research, University of Texas.

Taylor, F. W. (1911). *The principles of scientific management.* New York: Harper and Bros.

Tompkins, J. R. (2005). *Organization theory and public management.* Belmont, CA: Thomson Wadsworth.

Usher, A. P. (1920). *An introduction to the industrial history of England.* Boston: Harvard University Press.

Weick, K. E. (1979). *The social psychology of organizing* (2nd ed.). Boston: McGraw Hill.

Wood, J., Wallace, J., Zeffane, R., Champan, J., Fromholtz, M., & Morrison, V. (2004). *Organisational behaviour: A global perspective* (3rd ed., pp. 355–357). Queensland, Australia: John Wiley & Sons.

Zaleznik, A. (1977, May–June). Managers and leaders: Are they different? *Harvard Business Review.*

Zuboff, S. (1988). *In the age of the smart machine.* New York: Basic Books.

Index

For Product Safety Concerns and Information please contact our EU
representative GPSR@taylorandfrancis.com Taylor & Francis Verlag GmbH,
Kaufingerstraße 24, 80331 München, Germany

Printed and bound by CPI Group (UK) Ltd, Croydon, CR0 4YY
08/05/2025
01864479-0001